Nanomaterials for Energy Applications 2015

Topical Conference at the 2015 AIChE Annual Meeting

Salt Lake City, Utah, USA
8-13 November 2015

ISBN: 978-1-5108-1845-3

Printed from e-media with permission by:

Curran Associates, Inc.
57 Morehouse Lane
Red Hook, NY 12571

proceedings.com

Some format issues inherent in the e-media version may also appear in this print version.

Printed by Curran Associates, Inc. (2016)

AIChE
120 Wall Street, FL 23
New York, NY 10005-4020

Phone: (800) 242-4363
Fax: (203) 775-5177

www.aiche.org

Additional copies of this publication are available from:

Curran Associates, Inc.
57 Morehouse Lane
Red Hook, NY 12571 USA
Phone: 845-758-0400
Fax: 845-758-2634
Email: curran@proceedings.com
Web: www.proceedings.com

TABLE OF CONTENTS

(26a) Low-Temperature Approaches to Inorganic Photovoltaic Thin Films 1
Rainie D. Nelson, Umar H. Hamdeh, Matthew G. Panthani

(26b) Symmetry-Breaking in Light-Trapping Nanostructures on Silicon for Solar Photovoltaics 2
Sang Eon Han, Seok Jun Han, Swapnadip Ghosh, Tianhao Cai, Brittany R. Hoard, Sang M Han

(26c) Controlling Morphology of Photovoltaic Thin Films By Phase Transformation of Metastable Colloidal Nanocrystals 3
Ajay Singh, Gary K. Ong, Delia J. Milliron

(26d) Plasmonic Enhancement of Mesoporous Solar Cells with Shape Controlled Nanostructures 4
Rizia Bardhan

(26e) Effect of Uniquely Assembled Nanostructures on Photovoltaic Properties 5
Nurxat Nuraje

(26f) Applying the Chemistry of Amine-Thiol Mixtures for Solution-Processed CdTe Thin Films 6
Caleb Miskin, Robert W. Boyne, Rakesh Agrawal

(26g) Production of $Cu(InGa)(SeS)_2$ Thin Films Via Rapid Thermal Processing of Selenium-Capped Precursors 7
Robert J. Lovelett, William N. Shafarman, Robert W. Birkmire, Babatunde A. Ogunnaike

(80a) Metallization of DNA Origami to Form Thin, Electrically Conductive Nanowires 27
Bibek Uprety, John Harb

(80b) Novel Light-Activated Therapy for Mutli-Drug Resistant Pathogens 28
Samuel Goodman, Colleen Courtney, Anushree Chatterjee, Prashant Nagpal

(80c) Rapid Synthesis of Unilamellar Liposomes 29
P Sunthar, Sopan M Phapal

(80d) Encapsulation and Release of Hydrophilic Drug Molecules Via a Poly(lactic acid)-Montmorillorite Composite Micro/Nano-Particle System 30
Anna Song, Joung Sook Hong, Ilsoon Lee, Shaowen Ji

(80e) Biocompatibility of Lipid Coated Nanocomposites 31
Alexander Kelly, Robert Arnold, Allan David

(80f) Biomimetic Amphiphilic Polymers for Intracellular Therapeutic Delivery and Theranostic Applications 32
Rongjun Chen

(98a) Water Desalination By Shock Electrodialysis 33
Nancy Lu

(98b) Exergy Analysis of a Power Plant in Abu Dhabi (UAE) 34
Abdullah Alhosani

(98c) Insights into the Hydrothermal Stability of ZSM-5 Under Relevant Biomass Conversion Reaction Conditions 35
David W. Gardner

(98d) Effects of Season and Heating Mode on Ignition and Burning Behavior of Ten Species of LIVE FUEL Measured in a FLAT-Flame Burner System 36
Samantha Smith

(98e) Characterization and Particle Sizing of the Composition of E-Cigarette Aerosol 45
Jordan Berger

(98f) Ammonia Removal from Aquaculture Stocking Water 46
Martin C

(98g) Impact of Chemical Dopants and Passivation Schemes on Carbon Nanotube Sheet Conductivity 47
Colleen C. Lawlor

(98h) Îœfilms: Dynamics of Thin-Films Under Physiological Fluids and Shear Flow 48
Monica Torralba

(98i) Photocatalytic Methanol Reforming on TiO2 49
Katelyn Dagnall

(98j) Room Temperature Shape Memory Polymers 50
Heather Fairbairn

(159a) Suppression of Infrared Absorption in Nanostructured Metals By Controlling Faraday Inductance and Electron Path Length 51
Sang Eon Han

(159b) Plasmon-Enhanced Energy Transfer and Other Photophysical Effects in Doped-Lanthanide Nanocrystals 52
Qi-C. Sun, Prashant Nagpal
(159c) Quantum Dynamical Simulations of the Photoinduced Charge Transfer Process in Donor-Bridge-Acceptor 53
M. Belen Oviedo, Bryan Wong
(159d) Multiple Energy "Exciton-Shelves" in Quantum-Dot-DNA Nanobioelectronic Materials 54
Prashant Nagpal, Samuel Goodman
(159e) Integration of Photosystem I Proteins within Conductive Polymer Matrices Using Vapor Phase Techniques 55
Maxwell Robinson, Evan Gizzie, G. Kane Jennings, David Cliffel
(218a) Solar Evaporation Enhancment Using Floating Copper Oxide Deposited Cellulose Paper 56
Amin Yoosefi Booshehri, Rong Xu
(218b) Bottom-up Assembly of Metal Silicide Nanowires into Highly Efficient Bulk Thermoelectrics 57
Sreeram Vaddiraju, Yongmin Kang, Venkata Vasiraju
(218c) Omni-Thermoelectrics: Atomically Convertible p/n Nanowire Inks for Flexible Generators 58
Ayaskanta Sahu, Boris Russ, Miao Liu, Jason Forster, Nav Nidhi Rajput, Fan Yang, Raffaella Buonsanti, Chris Dames, Kristin Persson, Jeffrey Urban, Rachel Segalman
(218d) Computational Study of Thermal Transport in Si-Ge Nanostructures - Exploration of Phonon Scattering Contributions to Suppressed Conductivity 59
Yongjin Lee, Alexander Pak, Gyeong Hwang
(218e) A Comparative Study of Two Narrow Gap Semiconductors FeGa3 and FeSb2 60
Lianyang Dong, Theo Siegrist
(286a) Understanding the Configuration-Mechanical Stability Relationships for Si-CNT Heterostructured Anodes for Li-Ion Battery: A Computational Study 61
Sameer Damle, Siladitya Pal, Spandan Maiti, Prashant N. Kumta
(286b) A Fully Integrated, Efficient and Stable Solar-Driven Water-Splitting Prototype 63
Chengxiang Xiang
(286c) Lithium Sulfide Cathodes Via Aerosol Spray Pyrolysis 64
Noam Hart, Juchen Guo
(286d) The Role of Polymer Composite Binder on Mechanics and Performance of Lithium Ion Battery Electrodes 65
Thomas Humplik, Anne M. Grillet, Dave A. Barringer, Emily K. Stirrup, Kevin N. Long, Hector Mendoza, Scott A. Roberts, Chelsea Snyder, Christopher A. Apblett, Kyle R. Fenton
(286e) Electropolymerized Polyaniline/Manganese Iron Oxide Hybrids with Enhanced Electrochemical Energy Storage and Color Switching Response 66
Zhanhu Guo, Yiran Wang, Huige Wei, Guo Jiang, Suying Wei
(286f) Performance Metrics and Design Principles for Cost-Effective Separators in Non-Aqueous Redox Flow Batteries 67
Liang Su, Robert Darling, Kevin Gallagher, Wei Xie, Fikile Brushett
(286g) The Performance of Structured High-Capacity Si Anodes for Lithium-Ion Batteries 68
Juichin Fan, Lawrence Barrett, Sydney Palmer, Robert C. Davis, Richard R. Vanfleet, John Harb
(286h) Graphene Oxide Based Foams for Lithium Ion Batteries 76
Kurt B. Smith, M. Silvina Tomassone
(286i) Solubility Product Observations in Formation of Active LiFePO4 Cathodic Material for Secondary Energy Storage Application 77
Darren W. Kwee, Alfredo A. Martinez-Morales
(286j) Understanding Lithiation Mechanisms in Silicon-Based Nanomaterials from First Principles 78
Chiayun Chou, Gyeong Hwang
(306a) Novel Centrifugation-Assisted Preparation (CAP) of Additive-Free Carbon-Decorated Fe3O4 Electrodes with Superior Electrochemical Performances for Lithium-Ion Batteries 79
Jian Zhu, K. Y. Simon Ng, Da Deng
(306b) Stacked SnS2/Graphene Nanocomposites with High Li+ Storage Capacity 81
Baihua Qu, Ge Ji, Bo Ding, Meihua Lu, Weixiang Chen, Jim Yang Lee
(306c) Gas-Assisted Electrospray for Ultrauniform, Ultrafast, Instantly-Dry, and Binder-Free Electrode Preparation 82
Ling Fei, Yong Lak Joo
(306e) SnO2/Carbon Nanotube–Infiltrated Ni Nanofoams As 3D Anodes for Enhanced Performance of Lithium-Ion Batteries 83
Marissa Follette, Daniel R. Huffman, Jennifer Carpena, Michael F. Durstock, Benji Maruyama, Placidus B. Amama

(306f) Synthesis of Layer-By-Layer Thick Mesoporous Titania Films with Vertically Oriented 2D-HCP Nanopores and Their Use in Lithium Ion Batteries As Negative Electrodes 84
Suraj Nagpure, Syed Z. Islam, Qinglin Zhang, Yang -Tse Cheng, Stephen E. Rankin
(306g) Engineered Ionic Diffusion Layers to Increase Rate Capability of NCA Cathode in Lithium-Ion Cells 85
Kevin Dahlberg, Debasish Mohanty, Vishal Mahajan, Myongjai Lee, Lisa Stevenson, Joel Stanley, David King, David Wood, Fabio Albano, Subhash Dhar
(361a) Relationship of Graphene Microstructrue and Ultracapacitive Property 86
Xiong Zhiyuan
(361b) N-Doped Material for Supercapacitor from Lignin Biomass 93
Muslum Demir, Ahmed A. Farghaly, Maryanne M Collinson, Burak Aksoy, Harry T. Cullinan, Hani M El-kaderi, Timur Islamoglu, Ram B. Gupta
(361c) Nanostructured Surfactant Ionic Liquids with Unusually High Capacitances for High-Temperature Flexible Supercapacitors 94
Xianwen Mao, Paul Brown, Margarida Costa Gomes, T. Alan Hatton
(361d) Spinel Decorated Aligned Carbon Nanotube Arrays As Supercapacitor Electrodes 95
Moses Oguntoye, Noshir S. Pesika
(361e) Thin Film Manganese Oxide Electrodes for Batteries and Supercapacitors As Grown By Atomic Layer Deposition 96
Matthias J. Young, Markus Neuber, Christopher D. Hare, Hans-Dieter Schnabel, Charles B. Musgrave, Steven George
(361f) First-Principles-Based Analysis of the Potential Benefits of Graphene Oxide for Supercapacitor Applications 97
Alexander Pak, Gyeong Hwang
(414a) A Self-Supported and Long-Life Li-Se Battery Cathode Enabled By 3D Mesoporous Carbon/Graphene Hierarchical Architecture 98
Kai Han
(414b) Synergy Between Different Graphene Structures in Layer-By-Layer Assembly for Li-Air Battery Cathodes 99
Jangwoo Kim, Yong L. Joo
(414c) Synthesis and Electrochemical Characterization of Graphene-Metal and Metal Oxide Nanocomposites 100
Donald Johnson, Ashley Ware, Paul Gorrell, David Baah, Jonathan C. Mbah, Nader Vahdat
(414d) Co-Continuous Composite Electrodes Derived from Bijels That Concurrently Deliver High Energy and Power 113
Jessica A. Witt, Daniel R. Mumm, Ali Mohraz
(414e) Hierarchical Carbon-Based Electrode Materials for Vanadium Redox Flow Batteries 114
Panagiotis Trogadas, Tobias Neville, Dan Brett, Paul Shearing, Marc-Olivier Coppens
(414f) Artificial Photosystem I and II: Highly Selective Solar Fuels and Tandem Photocatalysis 115
Yuchen Ding, Prashant Nagpal
(414g) Charge-Storage Mechanisms for High Surface Area Carbides and Nitrides 116
Abdoulaye Djire, Jason Siegel, Lilin He, Alice E. S. Sleightholme, Saemin Choi, Paul G Rasmussen, Levi T. Thompson
(486a) Micro Structural and Electrical Properties of Ceria-Yttria Stabilized Zirconia Nanocomposites 119
Alka Gupta, Kantesh Balani
(486b) Nanostructured Robust Cobalt Alloy Based Anode Electro-Catalysts with Superior Electrochemical Activity for Proton Exchange Membrane Fuel Cells 120
Prasad P. Patel, Moni Kanchan Datta, Oleg Velikokhatnyi, Prashanth Jampani, Prashant N. Kumta
(486c) Pathways Towards Defect-Tolerate, Electrochemically Stable Solar-Hydrogen Membranes 123
Shu Hu, Ke Sun, Matthew Shaner, Michael Lichterman, P. Daniel Dapkus, Bruce S. Brunschwig, Nathan S. Lewis
(486d) PtCo/Coox Nanocomposites As Bifunctional Electrocatalysts for Oxygen Reduction and Evolution Reactions Synthesized Via Tandem Laser Ablation Synthesis in Solution-Galvanic Replacement Reactions 124
Sheng Hu
(486e) Activity of Electroless Deposited Transition Metal-Catalyst for the Alcoholysis of Ammonia Borane 125
Egwu E. Kalu, James A. Omoleye, Edith O. Onyeozili, Vincent Efeovbokhan
(548a) Monoclinic WO_3 Nanomultilayers with Preferentially Exposed (002) Facets for Photoelectrochemical Water Splitting 126
Jijie Zhang, Tuo Wang, Jinlong Gong
(548b) Transparent ALD-Grown Ta_2O_5 Protective Layers for Corrosion Vulnerable Photoanodes in Solar Water Splitting 130
Tuo Wang, Chengcheng Li, Jinlong Gong

(548c) Bimetallic Cu/Pd Nanoparticles As Low Temperature Sulfur-Tolerant WGS Catalysts 131
Vadim V. Guliants, SeongUk Yun
(548d) N2/Ar-Plasma Assisted Nitrogen Doping of Ordered Mesoporous TiO2 Thin Films for Water Splitting Photocatalysis 132
Syed Z. Islam, Allen Reed, Doo Young Kim, Stephen E. Rankin
(548e) TiO2/BaTiO3 Bi-Layered Approach Towards Photoelectrochemical Generation of Hydrogen Via Direct Splitting of Water 133
Shailja Sharma
Author Index

432868 Low-Temperature Approaches to Inorganic Photovoltaic Thin Films

Abstract:

Solution-based precursors are promising route to synthesizing thin film photovoltaic semiconductors which can be printed using low cost, high-throughput techniques such as roll-to-roll processing. A major limitation is the processing of these precursors to semiconductors; in order to achieve electronic and optoelectronic properties which are suitable for application in photovoltaics, high temperatures and harsh processing atmospheres are often required, which limits the benefits of solution-based precursors. In this presentation, we discuss some low-temperature approaches to synthesizing thin films at low temperature with excellent electronic properties. For example, we find that metastable wurtzite CdTe tetrapods undergo a phase transformation at 350°C and undergo sintering to produce grain sizes of several hundred nanometers. Additionally, thin films passivated with compositionally-matched chalcogenidomettalate clusters can have electron carrier mobilities in excess of 200 $cm^2 V^{-1} s^{-1}$. When integrated into photovoltaic devices, we find that the power conversion efficiencies greater than 12% along with evidence that much higher efficiencies could be achieved with optimization of contact layers.

429232 Symmetry-Breaking in Light-Trapping Nanostructures on Silicon for Solar Photovoltaics

Abstract:

In thin-film photovoltaics, highly absorptive materials are conventionally used. However, these materials have achieved efficiencies that are not comparable to those of thick crystalline silicon (c-Si) photovoltaics and, in some cases, suffer from their toxicity and low supply. A viable solution to these problems would be to use c-Si for thin-film photovoltaics. However, thin c-Si films absorb sunlight weakly because of its indirect band gap and strong light-trapping should be provided to achieve high efficiency. For thin-film photovoltaics, nanoscale structures are typically involved for light trapping because the film thickness becomes comparable to the wavelength of sun light. While diverse nanostructures have been studied to break the light-trapping limit of geometric optics, known as the Lambertian limit, highly efficient nanostructures that can be easily manufactured have not been demonstrated. We have previously predicted that symmetry-breaking in light-trapping periodic nanostructures on thin films can approach the Lambertian limit very closely. Herein, we will present how the systematic symmetry-lowering increases light-trapping in c-Si thin-film photovoltaics. We will demonstrate the experimental realization of such low-symmetry structures using simple wet etching methods on c-Si(100) wafers without any off-cut, tilt angle. Further, we will discuss the optical characterization of our fabricated structures on thin c-Si films.

425940 Controlling Morphology of Photovoltaic Thin Films By Phase Transformation of Metastable Colloidal Nanocrystals

Abstract:

Colloidal nanocrystals are promising solution processible precursors to thin films of complex semiconductors, including $CuZnSnS_4$ and $CuInSe_2$. Detrimental impurity phases and spatial fluctuations of the composition can be minimized by controlling stoichiometry during colloidal synthesis. However, the preparation of large-grained, electronic quality thin films from such nanocrystals has been challenging. We synthesize nanocrystals of these materials in a metastable hexagonal crystal phase and leverage their thermally driven conversion to the stable phase to engineer grain size and morphology in the resulting polycrystalline films. The activation energy for the phase transformation, determined by variable heating rate *in situ* X-ray diffraction, is found to depend strongly on nanocrystal shape. Using nanorods around 40 nm in length, we apply classical nucleation and growth principles to tune the thermal processing conditions to achieve large, micron scale grains in a matter of minutes. To exclude carbon impurities that can negatively impact optoelectronic properties, the same nanocrystals can be processed using inorganic chalcogenidometallates (ChaMs) as ligands. Upon heating, the ChaMs form a glassy matrix around each nanocrystal that inhibits nucleation of the stable phase (observed as a higher activation energy for nucleation), yet facilitates grain growth by coarsening. Interdiffusion ultimately results in alloyed thin films with grains that span their full thickness. Optical and electronic characterization of the resulting thin films shows promise for their application in thin film photovoltaic devices.

427513 Plasmonic Enhancement of Mesoporous Solar Cells with Shape Controlled Nanostructures

Abstract:

Plasmonic Enhancement of Mesoporous Solar Cells with Shape Controlled Nanostructures

Rizia Bardhan

Department of Chemical and Biomolecular Engineering,

Vanderbilt University, Nashville, TN 37235

Next generation mesoporous solar cells (MSCs) including dye-sensitized solar cells (DSSCs) and perovskite-sensitized solar cells (PSSCs) have rapidly emerged with efficiencies reported up to 13% for DSSCs and 20% for PSSCs. Current efforts to improve the performance of MSCs are focused on altering the photoanode morphology, and manipulating the sensitizer composition. In contrast to these approaches that often lead to incremental improvement tailored to each MSC system, plasmonic enhancement provides a universal route applicable to the whole family of MSCs that can significantly boost the optical absorption and carrier generation. Through small additions of metal nanoparticles to MSCs (< 2%), the amount of sensitizer required to achieve high efficiency can be drastically reduced enabling thin film architectures compatible with scalable nanomanufacturing routes including roll-to-roll processing and inkjet printing. Here we demonstrate enhanced light harvesting in both DSSCs and PSSCs by embedding shape-controlled Au nanostructures, and Au/Ag core/shell bimetallic nanostructures in the photoanodes. Plasmonic enhancement resulted in 25 – 35 % enhancement in DSSCs and >15% enhancement in PSSCs. Photocurrent and IPCE spectra revealed that device performance in MSCs is controlled by particle density of plasmonic nanostructures and monotonically decreases at high concentrations. Transient absorption pump-probe spectroscopy revealed that presence of metal nanostructures leads to enhanced exciton generation, and faster electron injection into TiO_2 conduction band before recombination can occur in the bulk, resulting in shorter exciton lifetime. This work shows that shape-controlled plasmonic nanostructures can be employed as a universal platform for enhanced light-trapping in a range of solar devices.

420359 Effect of Uniquely Assembled Nanostructures on Photovoltaic Properties

Abstract:

This study focuses on the energy transfer and electron transport in the photovoltaic device with uniquely designed nanostructures of an acceptor/donor energy system and an electron carrier delivery system. Furthermore, the effect of both the assembled nanostructures of acceptor/donor energy delivery and the different morphology of nanomaterials on photovoltaic performance are studied.

Förster resonance energy transfer between the xanthene dye (donor of energy) and a new polymethine dye (acceptor of energy) was studied on the surface of TiO_2 films in DSSCs. The sensitization of semiconductor films by the donor acceptor compound leads to a doubling of energy conversion efficiency. Measurements show that this is due to the widening of photo sensitivity of the cell in the blue region of the spectrum. Increasing the total energy absorbed leads to increasing the number of generated charge carriers at the interface of TiO_2 and the dyes.

Zinc Oxide nanoarrays with rod and sheet morphologies were fabricated and assembled for electron transport study of the dye sensitized solar cells (DSSCs). A photoluminescence spectra study indicated that nanosheets had more defect density compared to nanorods, which resulted in a lower open circuit photovoltage. Electrochemical impedance spectroscopy (EIS) and dye absorption amount analysis further explained the high short-circuit current density (J_{sc}) of the DSSC with ZnO nanosheet structure, based on the dye-loading amount, effective electron lifetime, and electron density.

418019 Applying the Chemistry of Amine-Thiol Mixtures for Solution-Processed CdTe Thin Films

Abstract:

Solution-processing promises to revolutionize the electronic industry by providing cost-effective and rapid fabrication of flexible electronic devices. Whereas traditional electronic devices are processed on hard substrates using slow and costly vacuum-based batch processing, solution-processing enables rapid, roll-to-roll manufacturing on plastic films and foils. Electronic devices on such substrates are desirable for wearable electronics and to apply conformally to any surface. Important applications of solution-processed electronics include photovoltaics, thermoelectrics, thin film transistors, batteries, and sensors.

Ultimately, it is desirable to combine the superior performance of inorganic electronic materials with the facile processing of organic solutions. Unfortunately, most inorganic salts that would be useful precursors are insoluble in many organic solvents compatible with roll-to-roll, solution-based techniques. We have overcome this challenge by developing a mixture of commonly available thiols and amines to dissolve a host of materials that are otherwise insoluble in either solvent by itself. The solvent system not only takes the form as pure amine-thiol mixtures, but also as mixtures such as thioglycolic acid-ethanolamine,[1] ethanolamine-thioacetamide,[2] and ammonium thioglycolate solutions[3] in water. These mixtures have been used with great success for nanoparticle synthesis,[4,5] photovoltaic absorber layers,[1,6,7] luminescent quantum dot films,[1] and other thin films[8]making it a very general solvent system for processing of inorganic thin films.

Here we present for the first time the fabrication of CdTe thin films via a solution-processed molecular precursor approach. We further elucidate the chemistry of the amine-thiol system using advanced tandem mass spectrometry techniques such as ESI and APCI to identify species in solution and verify their structure using quantum chemical calculations and modeling. In a typical preparation $CdCl_2$and Te are each dissolved in 1-propanethiol and ethylenediamine in concentrations ranging from 0.15 M to 0.22 M. ITO-coated or Mo-coated glass are then spin-coated with 375 μL of the solution. To obtain a film thickness of about 1 μm, eight to twenty-five coatings are applied, depending on the concentration and spin coating speed employed. Between each coating, the film is annealed on a hot plate at 500 °C to grow the CdTe phase. Grazing incidence x-ray diffraction reveals that pure CdTe can be recovered under optimized annealing conditions. A plane view and cross section of a CdTe film are given in Figure 1. However, we have found that very short annealing times (less than 2 mins) have the potential to produce a small amount of a segregated Te phase. Scanning electron microscopy coupled with energy dispersive x-ray spectroscopy (EDS) linescans has confirmed the presence of this phase and found that it is located at the surface of the film.

Modest efficiencies of about 0.2% are currently obtained from these films. One limitation we have found is significant chlorine residue remaining in the film, often in between grain boundaries. Cadmium acetylacetonate can be used to either partially or entirely replace $CdCl_2$ as a precursor to reduce the presence of chlorine, but films prepared in this manner have much smaller grains than films prepared with $CdCl_2$and have more voids. Further insight into the solution chemistry and grain growth will enable the removal of chlorine contamination and improvement of device performance.

References

(1) Tian, Q.; Wang, G.; Zhao, W.; Chen, Y.; Yang, Y.; Huang, L.; Pan, D. *Chem. Mater.* **2014**, *26*(10), 3098.

(2) Sun, Y.; Zhang, Y.; Wang, H.; Xie, M.; Zong, K.; Zheng, H.; Shu, Y.; Liu, J.; Yan, H.; Zhu, M.; Lau, W. *J. Mater. Chem. A* **2013**, *1*(23), 6880.

(3) Tian, Q.; Huang, L.; Zhao, W.; Yang, Y.; Wang, G.; Pan, D. *Green Chem.* **2015**, *17*(2), 1269.

(4) Yang, W.-C.; Miskin, C. K.; Hages, C. J.; Hanley, E. C.; Handwerker, C.; Stach, E. A.; Agrawal, R. *Chem. Mater.* **2014**, *26*(11), 3530.

(5) Walker, B. C.; Agrawal, R. *Chem. Commun. (Camb).* **2014**, *50*(61), 8331.

(6) Yang, Y.; Wang, G.; Zhao, W.; Tian, Q.; Huang, L.; Pan, D. *ACS Appl. Mater. Interfaces* **2014**, *7*(1), 460.

(7) Zhang, R.; Szczepaniak, S. M.; Carter, N. J.; Handwerker, C. A.; Agrawal, R. *Chem. Mater.* **2015**, *27*(6), 2114.

(8) Antunez, P. D.; Torelli, D. A.; Yang, F.; Rabu, F. A.; Lewis, N. S.; Brutchey, R. L. *Chem. Mater.* **2014**, *26* (19), 5444.

Production of $Cu(InGa)(SeS)_2$ absorber layers via rapid thermal processing of selenium-capped metal precursors for thin film solar cells

R.J. Lovelett*†, W.N. Shafarman†, R.W. Birkmire†, B.A. Ogunnaike*

University of Delaware
*Department of Chemical and Biomolecular Engineering
†Institute of Energy Conversion

November 8, 2015

Introduction

Solar energy is experiencing rapid growth

- 36% of all newly installed US electric capacity in 2014 is solar-based

Thin films offer advantages over dominant technology, crystalline silicon

- Simple, low-cost manufacturing
- $CuInGaSe_2$ (CIGS) record cell efficiency is 21.7%, best among thin films and greater than mc-Si

There are multiple viable manufacturing routes to produce CIGS modules

- Elemental coevaporation
- Precursor reaction/selenization

Introduction

Solar energy is experiencing rapid growth

- 36% of all newly installed US electric capacity in 2014 is solar-based

Thin films offer advantages over dominant technology, crystalline silicon

- Simple, low-cost manufacturing
- $CuInGaSe_2$ (CIGS) record cell efficiency is 21.7%, best among thin films and greater than mc-Si

There are multiple viable manufacturing routes to produce CIGS modules

- Elemental coevaporation
- **Precursor reaction/selenization**

CIGS Properties

CIGS is an inorganic, p-type, polycrystalline semiconductor

- Chalcopyrite crystal structure
- Direct band gap with high optical absorption
- Forms a **continuous solid solution** of $CuInSe_2$ and $CuGaSe_2$
- Band gap can be adjusted by In:Ga ratio or alloying with Ag, Al, S or other elements
  - **Band gap can be tuned** to match the solar spectrum
  - **Through-film gradient** in material affects electrical properties

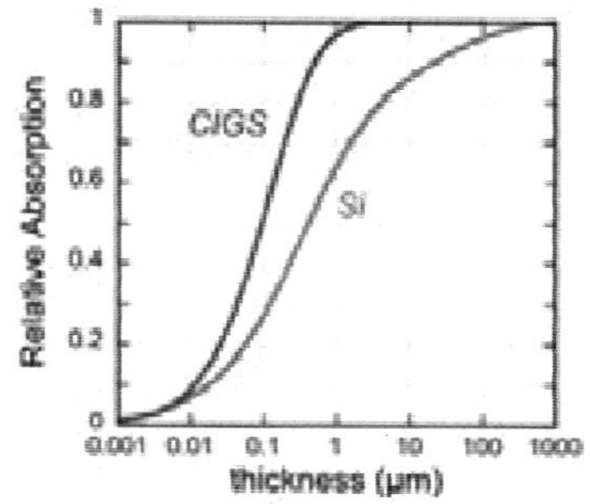

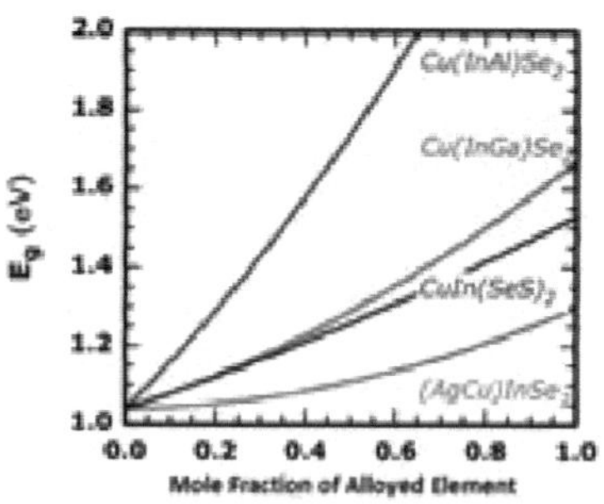

CIGS Solar Cells

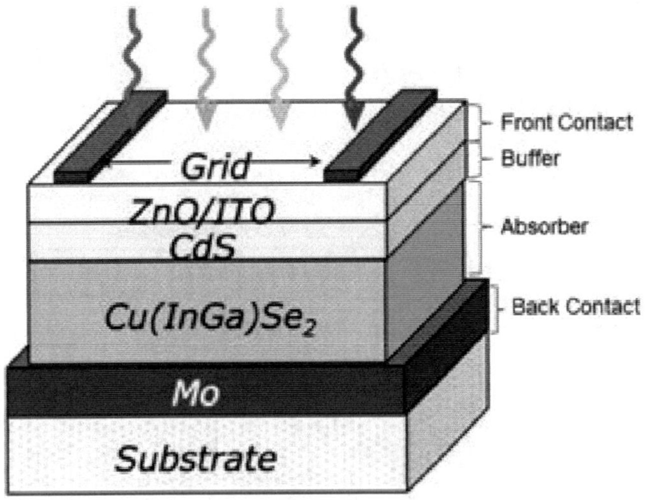

Device structure of a typical CIGS cell[1]

[1]Shafarman, Siebentritt, and Stolt, *Handbook of Photovoltaics*, 2011

CIGS Production by Precursor Reaction/Selenization

- Consists of two step process:
  1. Deposit Cu-In-Ga precursor
  2. React in H_2Se and/or H_2S gas
- Alternative approach: include selenium in precursor; react in H_2S
- Challenges:
  - Slow kinetics mean **low process throughput** (residence time ∼1 hour)
  - **Asymmetric reactions cause gradients** in through-film compositions: accumulation of Ga often seen at back contact

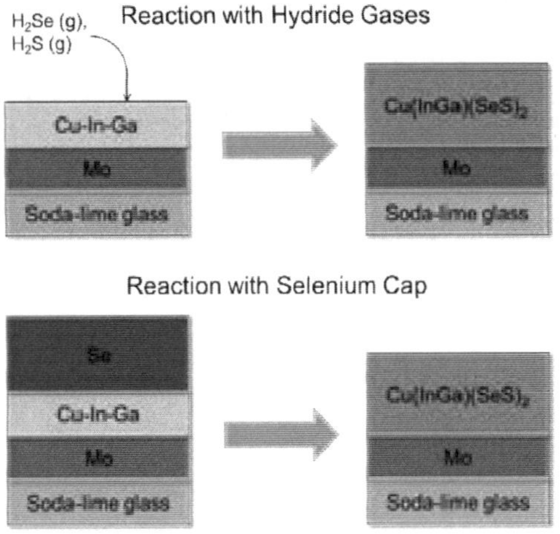

CIGS Production by Precursor Reaction/Selenization

- Consists of two step process:
  1. Deposit Cu-In-Ga precursor
  2. React in H_2Se and/or H_2S gas
- Alternative approach: include selenium in precursor; react in H_2S
- Challenges:
  - Slow kinetics mean **low process throughput** (residence time ~1 hour)
  - **Asymmetric reactions cause gradients** in through-film compositions: accumulation of Ga often seen at back contact

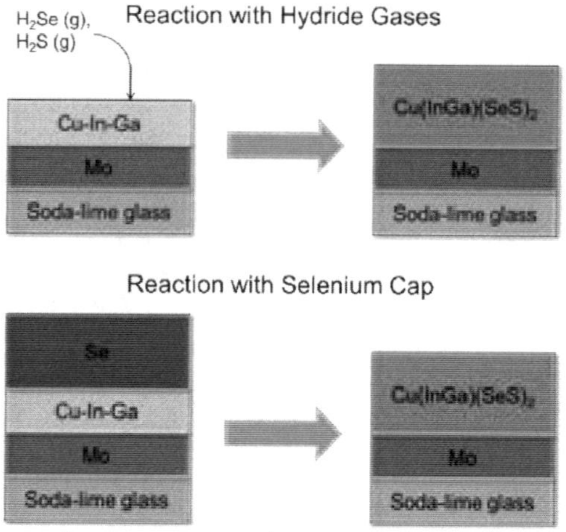

Approach: study reactions of Se-capped precursors in custom-built rapid thermal processing (RTP) reactor to develop fast process for absorber synthesis.

Motivation and Goals

- Selenization process promising method to produce high-efficiency modules with low cost and high throughput
- Slow kinetics and toxic gas handling limit throughput and technology deployment

Motivation and Goals

- Selenization process promising method to produce high-efficiency modules with low cost and high throughput
- Slow kinetics and toxic gas handling limit throughput and technology deployment

High temperature process using **Se-capped precursors** is viable alternative

- Process to be investigated with:
  1. Experimental studies: film and device properties
  2. Growth modeling and simulation

Motivation and Goals

- Selenization process promising method to produce high-efficiency modules with low cost and high throughput
- Slow kinetics and toxic gas handling limit throughput and technology deployment

High temperature process using **Se-capped precursors** is viable alternative

- Process to be investigated with:
  1. **Experimental studies: film** and device **properties**
  2. Growth modeling and simulation

Reaction with Se-Capped Precursor

CIGS Deposition Process:

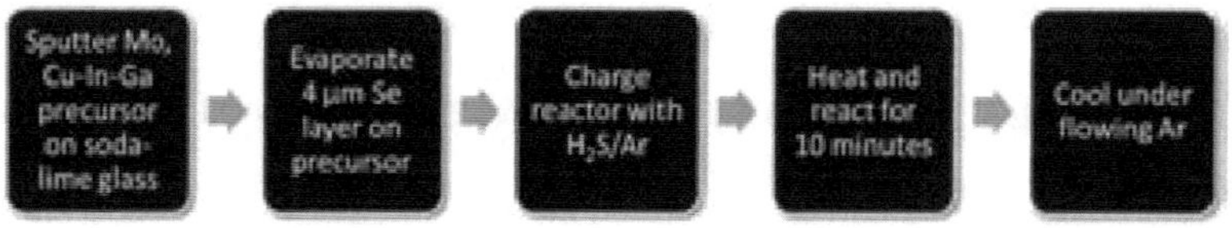

Approach: 2^3 Factorial Design
Factors:

- Temperature ramp rate (2–6 K/s)
- Dwell temperature (500–600 °C)
- H_2S concentration (1–10%)

Material Characterization/Responses:

- Elemental composition (EDX and XRF)
- Phase composition and analysis (XRD)

Film Properties

Process effects from EDX ($\sim 1/2$ depth of film sampled) results:

- S/(S+Se) and Cu/(In+Ga) increase with H_2S concentration
- Ga/(In+Ga) increases with temperature
- No effect of ramp rate within the range studied

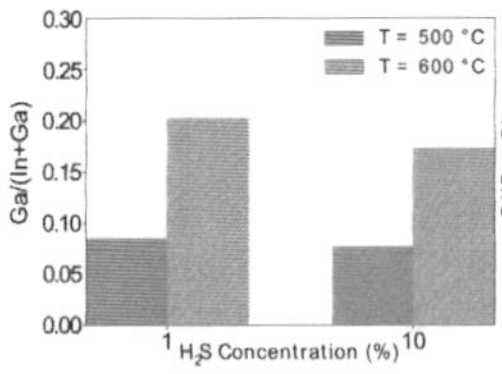

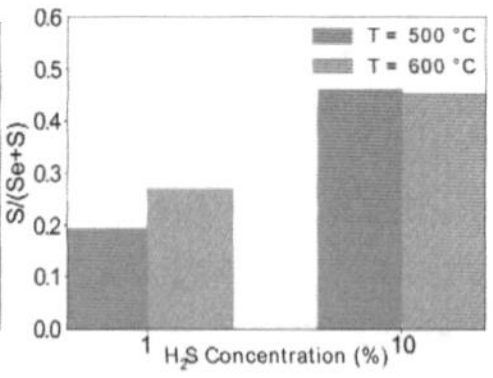

Film Properties

Process effects from EDX (~ 1/2 depth of film sampled) results:

- S/(S+Se) and Cu/(In+Ga) increase with H_2S concentration
- Ga/(In+Ga) increases with temperature
- No effect of ramp rate within the range studied

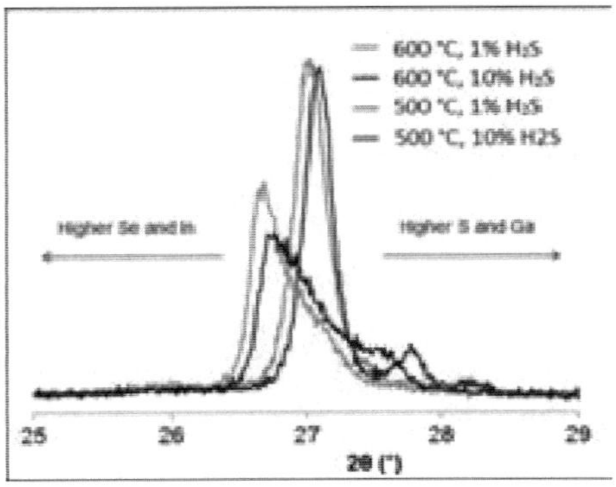

XRD patterns of the CIGS 112 peak of films produced with varying process conditions.

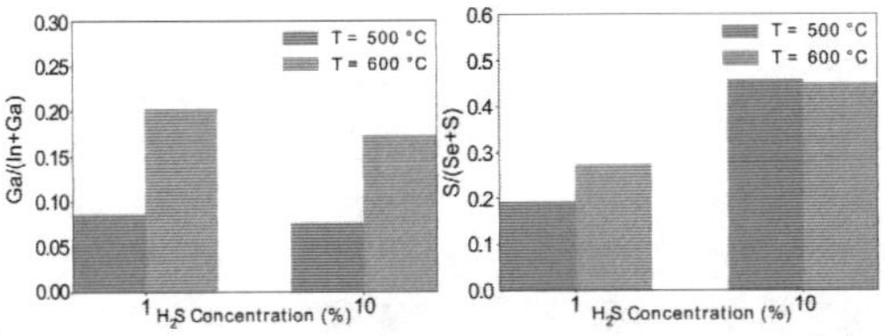

Time Evolution

Film reaches steady states within 5 minute reaction:

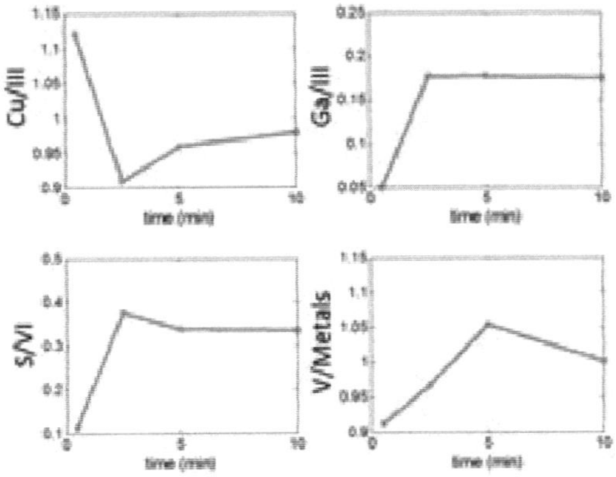

Time evolution of film composition

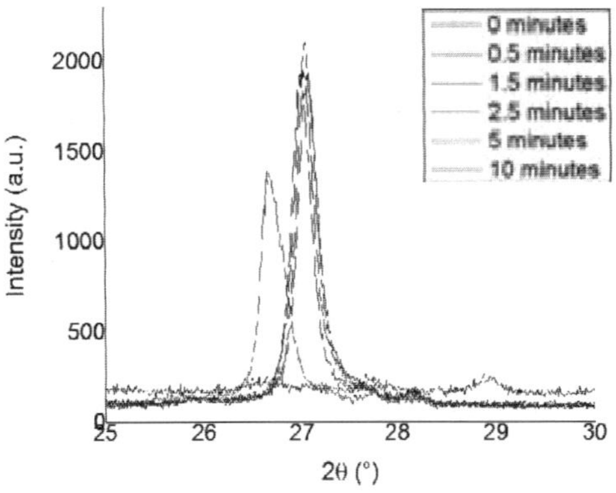

Time evolution of 112 peak

Motivation and Goals

- Selenization process promising method to produce high-efficiency modules with low cost and high throughput
- Slow kinetics and toxic gas handling limit throughput and technology deployment

High temperature process using **Se-capped precursors** is viable alternative

- Process to be investigated with:
  1. **Experimental studies:** film and **device properties**
  2. Growth modeling and simulation

Reaction with Se-Capped Precursor

CIGS Deposition Process:

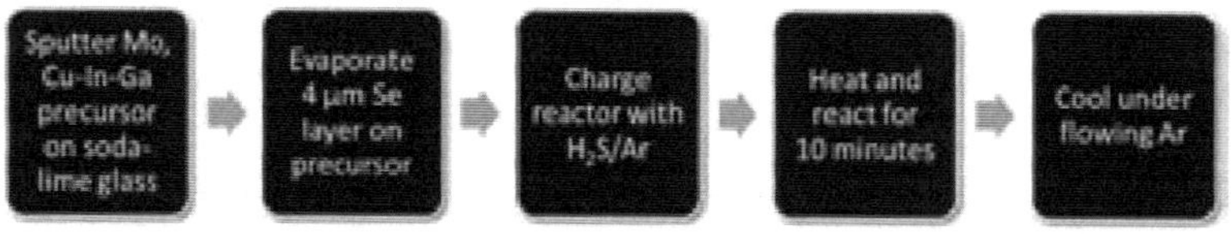

- Complete device synthesis: CdS CBD; sputter ZnO/ITO; Evaporate grids; mechanically scribe cells
- Measure JV and QE

Device Results: JV Curves

- Devices produced from CIGS from Se-capped process at varying H_2S and temperature.
- Devices from low temperature process shunted ($\eta \approx 0$). Small concentration unreacted phases likely.

Introduction Thin Film Characterization Device Results Stochastic Modeling Conclusions and Future Work

Device Results: JV Curves

- Devices produced from CIGS from Se-capped process at varying H_2S and temperature.
- Devices from low temperature process shunted ($\eta \approx 0$). Small concentration unreacted phases likely.

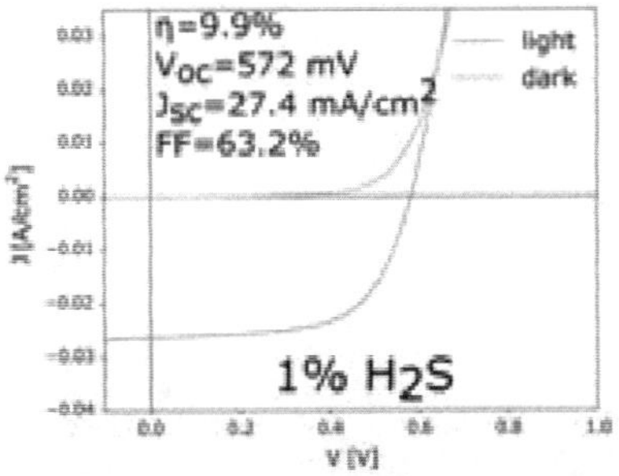

R.J. Lovelett, W.N. Shafarman, R.W. Birkmire, B.A. Ogunnaike University of Delaware
CIGS Production via Rapid Thermal Processing of Se-capped Precursors 13 of 22

Device Results: JV Curves

- Devices produced from CIGS from Se-capped process at varying H_2S and temperature.
- Devices from low temperature process shunted ($\eta \approx 0$). Small concentration unreacted phases likely.

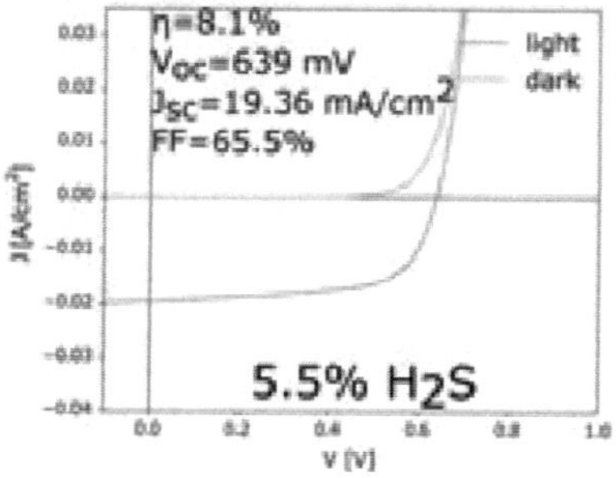

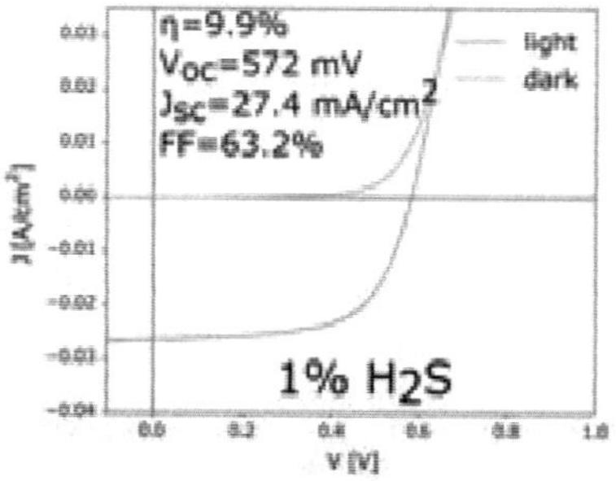

Introduction Thin Film Characterization **Device Results** Stochastic Modeling Conclusions and Future Work

Device Results: JV Curves

- Devices produced from CIGS from Se-capped process at varying H_2S and temperature.
- Devices from low temperature process shunted ($\eta \approx 0$). Small concentration unreacted phases likely.

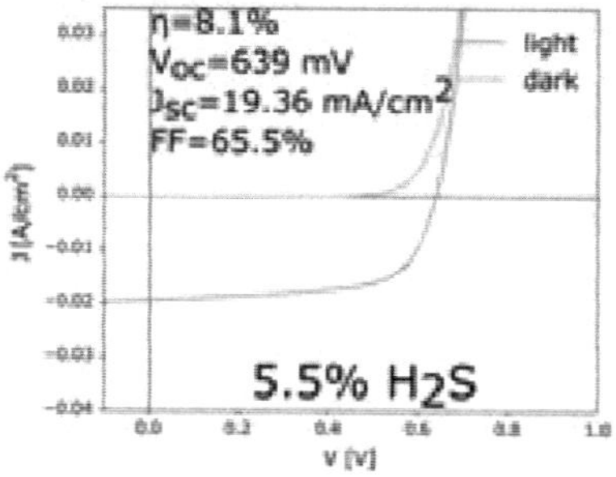

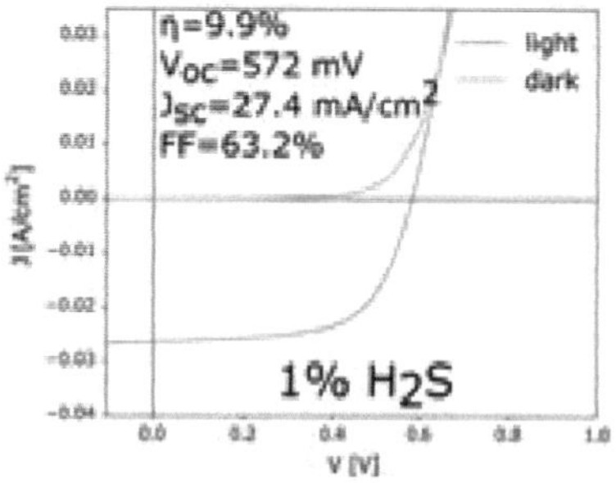

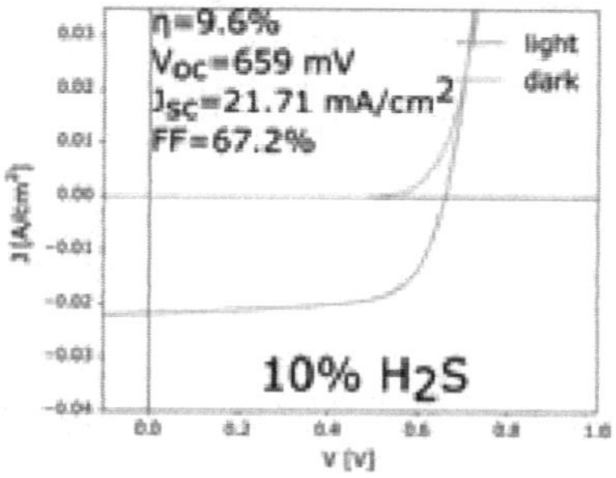

Device Results: QE Curves

- Quantum efficiency (QE) gives fraction of radiation converted to electricity at each wavelength
- Optical bandgap of device determined from inflection point in long-wavelength region

Introduction Thin Film Characterization Device Results Stochastic Modeling Conclusions and Future Work

Device Results: QE Curves

- Quantum efficiency (QE) gives fraction of radiation converted to electricity at each wavelength
- Optical bandgap of device determined from inflection point in long-wavelength region

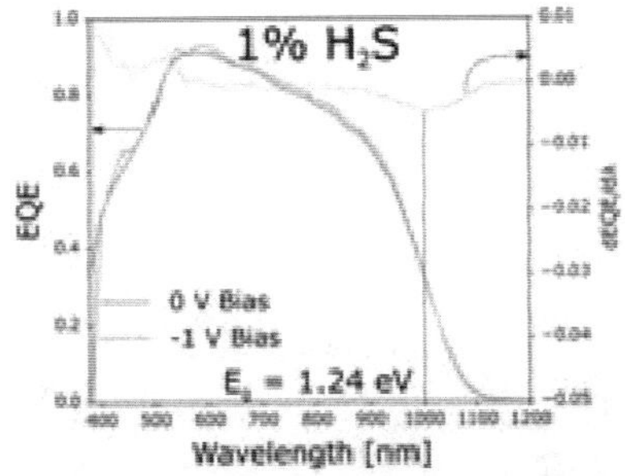

R.J. Lovelett, W.N. Shafarman, R.W. Birkmire, B.A. Ogunnaike University of Delaware
CIGS Production via Rapid Thermal Processing of Se-capped Precursors 14 of 22

Device Results: QE Curves

- Quantum efficiency (QE) gives fraction of radiation converted to electricity at each wavelength
- Optical bandgap of device determined from inflection point in long-wavelength region

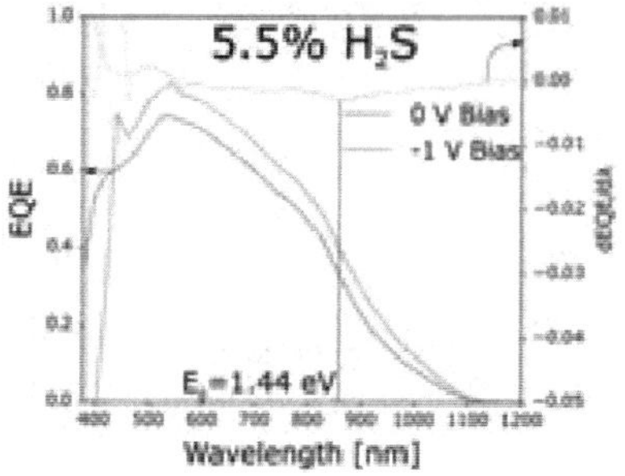

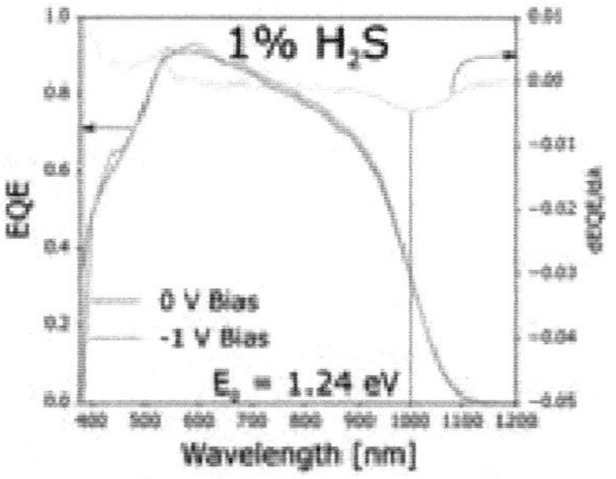

Introduction Thin Film Characterization Device Results Stochastic Modeling Conclusions and Future Work

Device Results: QE Curves

- Quantum efficiency (QE) gives fraction of radiation converted to electricity at each wavelength
- Optical bandgap of device determined from inflection point in long-wavelength region

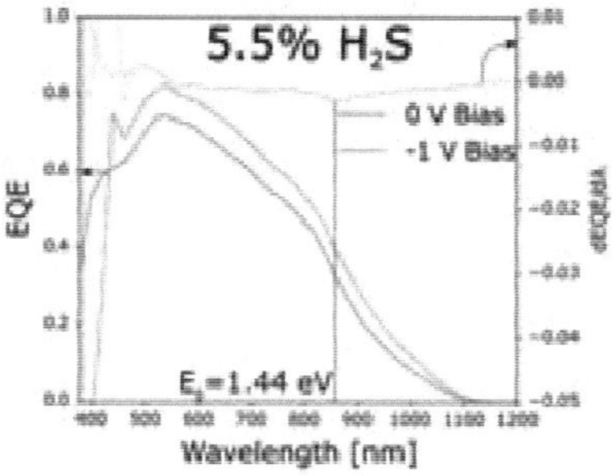

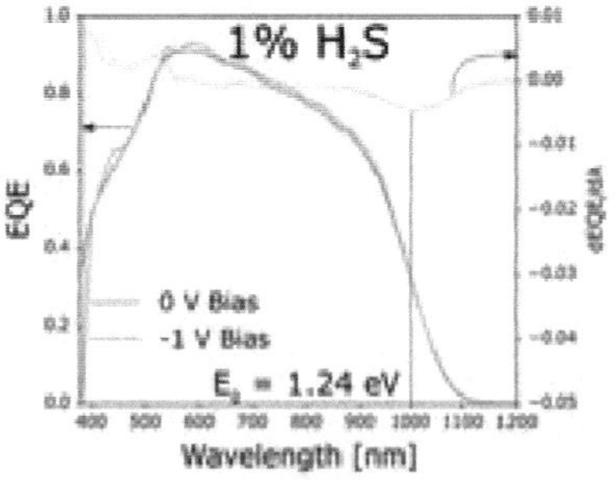

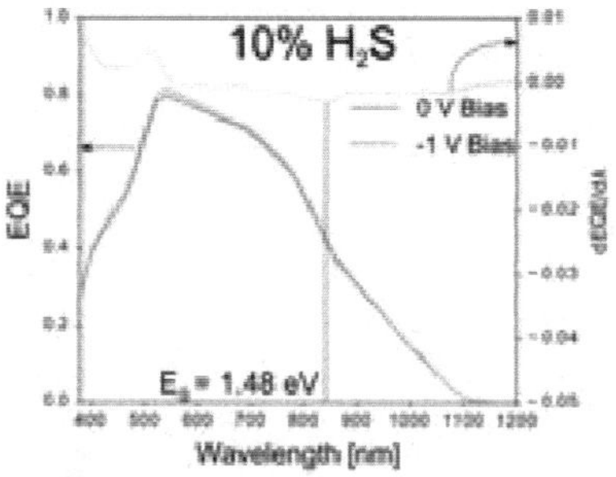

R.J. Lovelett, W.N. Shafarman, R.W. Birkmire, B.A. Ogunnaike University of Delaware
CIGS Production via Rapid Thermal Processing of Se-capped Precursors 14 of 22

Experimental Results: Summary and Analysis

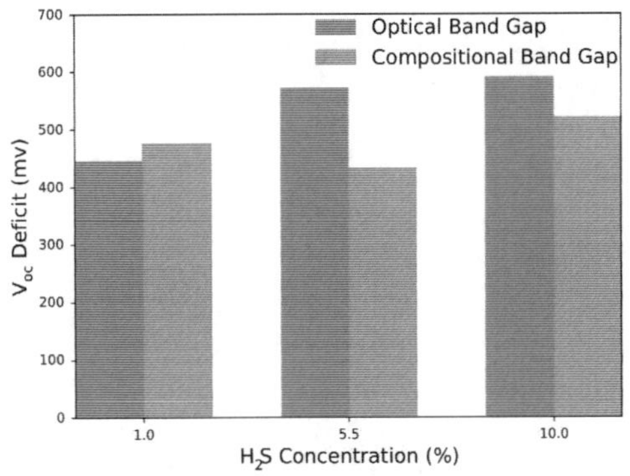

Voltage deficit vs. radiative limit using QE and compositional[1] band gaps

[1] Bär et al., *J. Appl. Phys.*, 2004

- Devices have well-behaved JV curves, but substantial losses compared to state-of-the-art.
- Losses in V_{OC} and FF; possibly high series resistance.
- Future Work: Identify source of V_{OC} deficit and voltage dependent QE with more thorough device analysis: JVT, capacitance profiling

Motivation and Goals

- Selenization process promising method to produce high-efficiency modules at low cost
- Slow kinetics and toxic gas handling limit throughput and technology deployment

High temperature process using **Se-capped precursors** is viable alternative

- Process to be investigated with:
  1. Experimental studies: film and device properties
  2. **Growth modeling and simulation**

Stochastic Model

Motivation: Explain experimental results and to assist with process improvements

Stochastic simulation provides a method to:

- Predict through-film composition
- Decouple and estimate kinetic and transport properties
- Access spatially-confined film properties (e.g., agglomerations)

Model used previously to simulate reactions in H_2Se and H_2S[2]

[2] R.J. Lovelett et al., *IEEE PVSC*, 2015

Introduction Thin Film Characterization Device Results Stochastic Modeling Conclusions and Future Work

Stochastic Model

Motivation: Explain experimental results and to assist with process improvements

Stochastic simulation provides a method to:

- **Predict through-film composition**
- Decouple and estimate kinetic and transport properties
- Access spatially-confined film properties (e.g., agglomerations)

Model used previously to simulate reactions in H_2Se and H_2S[2]

[2] R.J. Lovelett et al., *IEEE PVSC*, 2015

R.J. Lovelett, W.N. Shafarman, R.W. Birkmire, B.A. Ogunnaike University of Delaware
CIGS Production via Rapid Thermal Processing of Se-capped Precursors 17 of 22

Stochastic Simulation of Reaction

Reaction Mechanism/Lattice Events	
Adsorption/ Desorption	Selenium Adsorption = 0 Sulfur Adsorption Selenium Desorption Sulfur Desorption

Introduction | Thin Film Characterization | Device Results | Stochastic Modeling | Conclusions and Future Work

Stochastic Simulation of Reaction

Reaction Mechanism/Lattice Events	
Adsorption/ Desorption	Selenium Adsorption = 0 Sulfur Adsorption Selenium Desorption Sulfur Desorption
Binary Selenide/ Sulfide Formation	$CuIn + 2\,Se \longrightarrow CuSe + InSe$ $CuGa + 2\,Se \longrightarrow CuSe + GaSe$ $CuIn + 2\,S \longrightarrow CuS + InS$ $CuGa + 2\,S \longrightarrow CuS + GaS$

R.J. Lovelett, W.N. Shafarman, R.W. Birkmire, B.A. Ogunnaike | University of Delaware
CIGS Production via Rapid Thermal Processing of Se-capped Precursors | 18 of 22

Stochastic Simulation of Reaction

Reaction Mechanism/Lattice Events	
Adsorption/ Desorption	Selenium Adsorption = 0
	Sulfur Adsorption
	Selenium Desorption
	Sulfur Desorption
Binary Selenide/ Sulfide Formation	CuIn + 2 Se $\longrightarrow$ CuSe + InSe
	CuGa + 2 Se $\longrightarrow$ CuSe + GaSe
	CuIn + 2 S $\longrightarrow$ CuS + InS
	CuGa + 2 S $\longrightarrow$ CuS + GaS
Chalcopyrite Formation	CuSe + InSe $\longrightarrow$ $CuInSe_2$
	CuSe + GaSe $\longrightarrow$ $CuGaSe_2$
	CuS + InS $\longrightarrow$ $CuInS_2$
	CuS + GaS $\longrightarrow$ $CuGaS_2$
	CuS + InSe $\longrightarrow$ 0.5 $CuInSe_2$ + 0.5 $CuInS_2$
	CuS + GaSe $\longrightarrow$ 0.5 $CuGaSe_2$ + 0.5 $CuGaS_2$
	CuSe + InS $\longrightarrow$ 0.5 $CuInSe_2$ + 0.5 $CuInS_2$
	CuSe + GaS $\longrightarrow$ 0.5 $CuGaSe_2$ + 0.5 $CuGaS_2$

Introduction | Thin Film Characterization | Device Results | Stochastic Modeling | Conclusions and Future Work

Stochastic Simulation of Reaction

Reaction Mechanism/Lattice Events	
Adsorption/ Desorption	Selenium Adsorption = 0
	Sulfur Adsorption
	Selenium Desorption
	Sulfur Desorption
Binary Selenide/ Sulfide Formation	CuIn + 2 Se $\longrightarrow$ CuSe + InSe
	CuGa + 2 Se $\longrightarrow$ CuSe + GaSe
	CuIn + 2 S $\longrightarrow$ CuS + InS
	CuGa + 2 S $\longrightarrow$ CuS + GaS
Chalcopyrite Formation	CuSe + InSe $\longrightarrow$ $CuInSe_2$
	CuSe + GaSe $\longrightarrow$ $CuGaSe_2$
	CuS + InS $\longrightarrow$ $CuInS_2$
	CuS + GaS $\longrightarrow$ $CuGaS_2$
	CuS + InSe $\longrightarrow$ 0.5 $CuInSe_2$ + 0.5 $CuInS_2$
	CuS + GaSe $\longrightarrow$ 0.5 $CuGaSe_2$ + 0.5 $CuGaS_2$
	CuSe + InS $\longrightarrow$ 0.5 $CuInSe_2$ + 0.5 $CuInS_2$
	CuSe + GaS $\longrightarrow$ 0.5 $CuGaSe_2$ + 0.5 $CuGaS_2$
Diffusion	Any pair of species

R.J. Lovelett, W.N. Shafarman, R.W. Birkmire, B.A. Ogunnaike | University of Delaware
CIGS Production via Rapid Thermal Processing of Se-capped Precursors | 18 of 22

Stochastic Simulation of Reaction

Reaction Mechanism/Lattice Events	
Adsorption/ Desorption	Selenium Adsorption = 0 Sulfur Adsorption Selenium Desorption Sulfur Desorption
Binary Selenide/ Sulfide Formation	$CuIn + 2\,Se \longrightarrow CuSe + InSe$ $CuGa + 2\,Se \longrightarrow CuSe + GaSe$ $CuIn + 2\,S \longrightarrow CuS + InS$ $CuGa + 2\,S \longrightarrow CuS + GaS$
Chalcopyrite Formation	$CuSe + InSe \longrightarrow CuInSe_2$ $CuSe + GaSe \longrightarrow CuGaSe_2$ $CuS + InS \longrightarrow CuInS_2$ $CuS + GaS \longrightarrow CuGaS_2$ $CuS + InSe \longrightarrow 0.5\,CuInSe_2 + 0.5\,CuInS_2$ $CuS + GaSe \longrightarrow 0.5\,CuGaSe_2 + 0.5\,CuGaS_2$ $CuSe + InS \longrightarrow 0.5\,CuInSe_2 + 0.5\,CuInS_2$ $CuSe + GaS \longrightarrow 0.5\,CuGaSe_2 + 0.5\,CuGaS_2$
Diffusion	Any pair of species

- Discretize film into a square lattice of finite volume elements
- Each lattice element is phase pure
- Time evolution of film takes place in discrete lattice events

Introduction | Thin Film Characterization | Device Results | Stochastic Modeling | Conclusions and Future Work

Stochastic Simulation of Reaction

Reaction Mechanism/Lattice Events	
Adsorption/ Desorption	Selenium Adsorption = 0 Sulfur Adsorption Selenium Desorption Sulfur Desorption
Binary Selenide/ Sulfide Formation	$CuIn + 2\,Se \longrightarrow CuSe + InSe$ $CuGa + 2\,Se \longrightarrow CuSe + GaSe$ $CuIn + 2\,S \longrightarrow CuS + InS$ $CuGa + 2\,S \longrightarrow CuS + GaS$
Chalcopyrite Formation	$CuSe + InSe \longrightarrow CuInSe_2$ $CuSe + GaSe \longrightarrow CuGaSe_2$ $CuS + InS \longrightarrow CuInS_2$ $CuS + GaS \longrightarrow CuGaS_2$ $CuS + InSe \longrightarrow 0.5\,CuInSe_2 + 0.5\,CuInS_2$ $CuS + GaSe \longrightarrow 0.5\,CuGaSe_2 + 0.5\,CuGaS_2$ $CuSe + InS \longrightarrow 0.5\,CuInSe_2 + 0.5\,CuInS_2$ $CuSe + GaS \longrightarrow 0.5\,CuGaSe_2 + 0.5\,CuGaS_2$
Diffusion	Any pair of species

- Discretize film into a square lattice of finite volume elements
- Each lattice element is phase pure
- Time evolution of film takes place in discrete lattice events
- Two random variables, that are functions of lattice state and **propensity constants**, are chosen at each time step
  - Event Propensity: $a_i = p_i N_i$
  - Event Probability: $P_i = \frac{a_i}{\Sigma a_j}$
  - Event Time PDF: $f(\tau) = \frac{1}{\Sigma a_i} \exp\left(-\frac{\tau}{\Sigma a_i}\right), \tau \geq 0$

R.J. Lovelett, W.N. Shafarman, R.W. Birkmire, B.A. Ogunnaike | University of Delaware
CIGS Production via Rapid Thermal Processing of Se-capped Precursors | 18 of 22

Model Results

Higher sulfur concentration moves band gap gradient closer to front of film

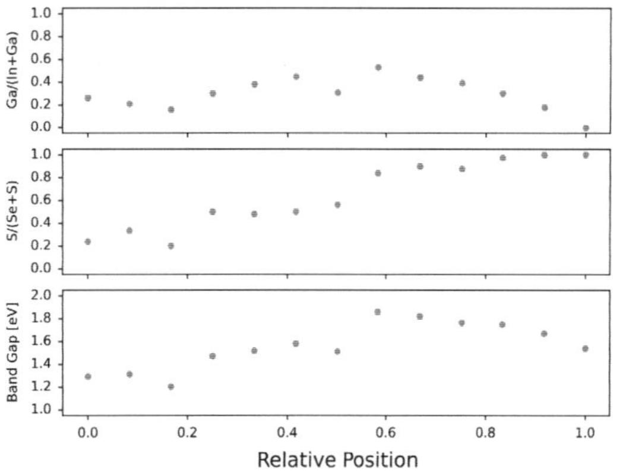

Profiles for low sulfur.

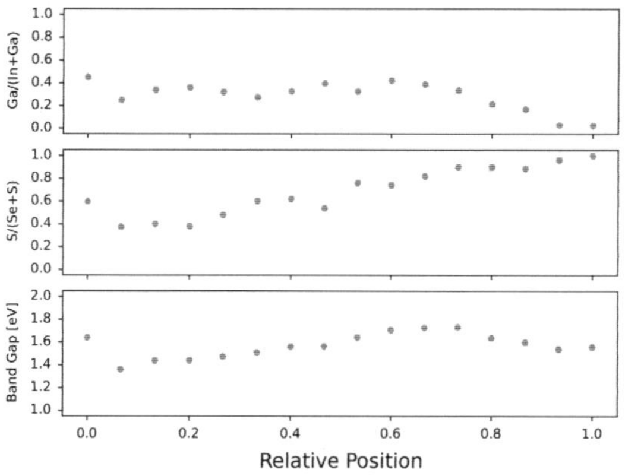

Profiles for high sulfur.

Conclusions and Future Work

Experimental Results:

- Films produced in <10 minutes at 600 °C

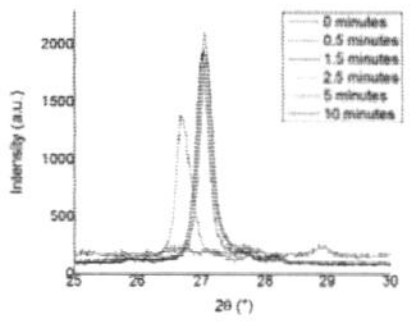

Conclusions and Future Work

Experimental Results:

- Films produced in <10 minutes at 600 °C
- $\eta \approx 10\%$ devices obtained with range of band gaps (varied S incorporation)
- Processing conditions affect the band gap and band gap gradient
- Next step: more thorough electrical characterization

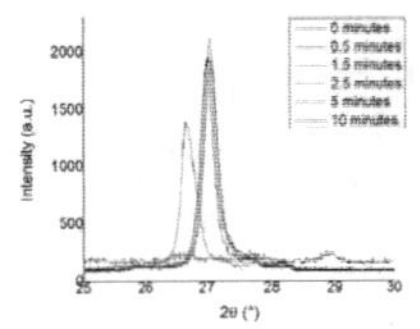

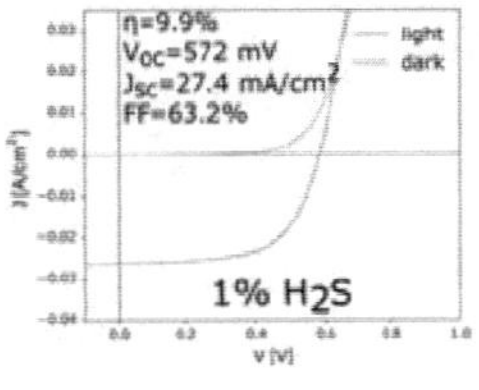

Conclusions and Future Work

Experimental Results:

- Films produced in <10 minutes at 600 °C
- $\eta \approx 10\%$ devices obtained with range of band gaps (varied S incorporation)
- Processing conditions affect the band gap and band gap gradient
- Next step: more thorough electrical characterization

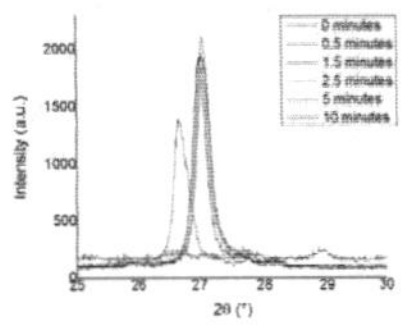

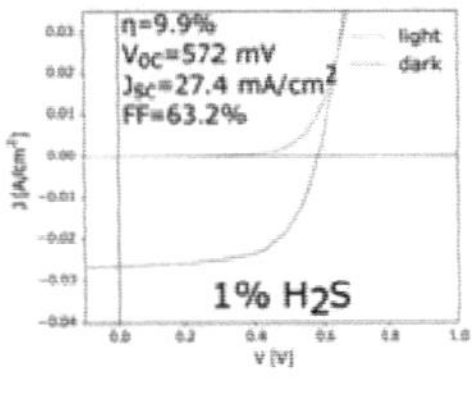

Modeling Results:

- Simulation results suggest mechanism for band gap grading

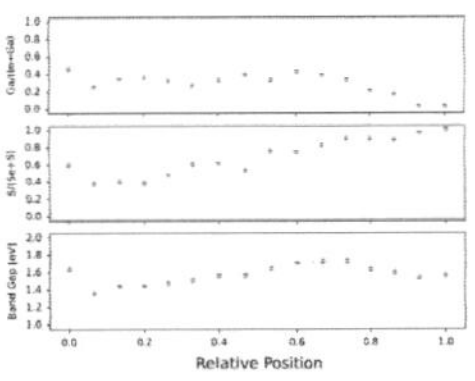

Acknowledgments

- Ogunnaike Research Group
  - Daniel Cook
  - Dr. Qian Gou
  - Dr. Jacob McGill
  - Dr. Lijia Luo
  - James Park
  - Xin Peng
  - Devesh Radhakrishnan
  - Dr. Melissa St. Amand
  - Chia-Hung Tsai
  - Dr. Zachary Whiteman
- Institute of Energy Conversion
  - Dr. Dominik Berg
  - Wayne Buchanan
  - John Elliot
  - Shannon Fields
  - Kevin Hart
  - Dr. Gregory Hanket
  - Brian McCandless
  - Sina Soltonmohammad

This material is based upon work primarily supported by the National Science Foundation (NSF) and the Department of Energy (DOE) under NSF CA No. EEC-1041895. Any opinions, findings and conclusions or recommendations expressed in this material are those of the author(s) and do not necessarily reflect those of NSF or DOE.

Production of $Cu(InGa)(SeS)_2$ absorber layers via rapid thermal processing of selenium-capped metal precursors for thin film solar cells

R.J. Lovelett*†, W.N. Shafarman†, R.W. Birkmire†, B.A. Ogunnaike*

University of Delaware
*Department of Chemical and Biomolecular Engineering
†Institute of Energy Conversion

November 8, 2015

426934 Metallization of DNA Origami to Form Thin, Electrically Conductive Nanowires

Abstract:

Bottom-up nanotechnology, which enables construction of complex structures from molecular building blocks, is seen as a promising technology to develop future generations of electronic devices. The bottom-up approach takes advantage of molecular self-assembly to create useful architectures of nanometer dimensions. DNA, with its small size, functional groups and complementary base pairs, is a powerful template for nanodevice fabrication via bottom-up assembly. Recent advances in the field of "DNA origami" have enabled fabrication of nanostructures by folding single-stranded DNA into a variety of 2- and 3-D shapes. However, electrically conductive structures are required for electronic applications.

In this study, we report the fabrication of electrically conductive DNA origami-templated nanowires with the use of gold nanorods (Au NR's) as a step towards DNA nanodevices. Our process begins with the seeding step where Au NR's electrostatically bind to the DNA origami backbone. The seeds are then connected by electroless deposition of gold to make continuous wires. The use of nanorods as seeds provides two key advantages over previous work with particulate seeds. First, the surfactant-enabled anisotropic growth used to initially fabricate the nanorods appears to persist through the electroless plating process on the DNA template. This permits improved control of the width of the final metallized structure since growth in the axial direction is preferred. The average final width of our Au-metallized DNA wires was 11.5 nm (n=30) with a standard deviation of 2.1 nm. We were able to obtain a 60% yield of continuous ~400 nm long DNA wires. In doing so, gaps between seed particles ranging from 10.6 nm on average to as wide as 25 nm were successfully filled. The second advantage of nanorods is the reduction in the number of connection points between seed particles by a factor of three relative to that observed for 5 nm seed particles. The reduced number of connection points is expected to increase the conductivity of the wires. Electrical characterization of the DNA wires is underway and complete current/voltage measurements will be presented.

In summary, we have demonstrated a metallization process to fabricate thin, high aspect ratio, conductive wires on DNA-origami templates with the use of Au NR's and electroless plating. Continuing efforts seek to use local functionalization to enable site-specific metallization of DNA-origami templates with Au NR's. The thin nanowires produced in this study will be interfaced with appropriate semiconducting elements to help enable fabrication of self-assembled nanoscale electronic circuits.

References:

1. Uprety, B.; Gates, E. P.; Geng, Y.; Woolley, A. T.; Harb, J. N. Site-Specific Metallization of Multiple Metals on a Single DNA Origami Template. *Langmuir* **2014**, 30, 1134-1141.
2. Gates, E.P.; Jensen, J.K.; Harb, J.N.; Woolley, A.T. Optimizing Gold Nanoparticle Seeding Density on DNA Origami. *RSC Advances* **2015**, 5, 8134-8141.

429915 Novel Light-Activated Therapy for Mutli-Drug Resistant Pathogens

Abstract:

Current design of therapeutics (drugs and nanoparticle-based therapies) relies heavily on a spatially selective drug-delivery system (colloids and vesicles coated with desired antibodies or proteins) due to non-specificity of the drugs or metal nanoparticles. Typically, these non-specific drugs once spatially delivered kill the targeted cells (e.g. tumor or pathogen) as well as healthy cells in close proximity (including metal nanopartciles used for cancer therapy). Further complexities in non-specific binding and incorrect delivery can cause undesirable side-effects. Here we present a novel nanoparticle-based therapy where the (non-metallic) nanoparticles specifically target selective cell phenotypes, based on their biochemical potentials. Using a specific light induced targeting mechanism, these light-activated reactive species (LARS) have the capability to target and kill specific-pathogens on illumination with visible light, without affecting the growth of other cells. While the growth of selected cells and the targeted pathogens is unaffected in dark, we selectively designed different LARS species which can either kill or proliferate the growth of selected target cells. This is demonstrated when targeting Escherichia coli cells, HEK293 mammalian cells, as well as, a co-culture of E. coli and HEK293 cells. Furthermore, within the same strain of E.coli, we find that while lab strain DH5Î± is susceptible to LARS, clinically isolated multi-drug resistant clinical strain of E. coli is not; likely due to presence of a non-specific antibiotic resistance mechanism. The ability to tune the biochemical potential of nanoparticles allows for selective cell phenotypes over a range of organisms, and has potential applications in antimicrobials and targeted organism-specific therapeutics.

422802 Rapid Synthesis of Unilamellar Liposomes

Abstract:

Liposomes present as a versatile drug carrying vehicle, capable of loading hydrophobic and hydrophilic drugs, bio-compatibility, long-circulation, wide mono-disperse size range (from 10 nm to 10s of microns), and targeting specific tissues. Several methods of liposome synthesis are being employed that result in production of unilamellar vesicles in a laboratory time scale of minutes. These fast methods employ external driving to achieve a monodisperse population. We present an alternate method of mono-disperse unilamellar liposome synthesis that results from spontaneous self-assembly of lipids in a solution (and therefore does not require external driving), and which is rapid. The lipids self-assemble when the solvent in which they are dissolved is rapidly replaced by a non-solvent (water) in a stationary phase interdiffusion. There has also been no attempt to resonably compare the synthesis times for various methods of liposome production. We provide estimates of the time of synthesis of a single liposome for various methods known in the literature, and show that the present method is among the fastest of all the methods, and is much faster than earlier methods of spontaneous assembly. The method is presently carried out in a batch mode, and requires device and process design to make it semi-batch or continuous.

426317 Encapsulation and Release of Hydrophilic Drug Molecules Via a Poly(lactic acid)-Montmorillorite Composite Micro/Nano-Particle System

Abstract:

Poly (lactic acid) (PLA) is one of the most widely used materials to fabricate biodegradable micro/nanoparticle drug delivery carriers. However, its hydrophobicity limits its application in the encapsulation of hydrophilic drug molecules, like nisin. In our group, a novel dynamic encapsulation process via a single emulsion method has been developed to encapsulate hydrophilic molecules within hydrophobic PLA micro/nano-particles. The uniqueness of this process is the emulsification of PLA solution in the water-glycerol medium under high viscous turbulent shear flow, where water-in-oil-water (W/O/W) droplets formed via one single emulsion step with the high temperature (HT) (60) condition, and oil-in-water (O/W) droplets under the room temperature (RT) (25) condition. The formation of W/O/W droplets and O/W droplets is reversible between the HT and RT condition, which can be controlled by a cooling circulating system. SEM results showed that open hollow microparticles were obtained under HT condition while nanospheres were prepared under RT condition. By increasing the repeating times of reversible droplets formation, more hydrophilic drug molecules can be encapsulated into the PLA particle matrix. The encapsulation efficiency and release profile will be tested on a series of samples with different repeating times. The addition of montmorillorite (MMT) to PLA is to improve the strength and the biocompatibility of polymer particles. The SEM results showed that more closed hollow microparticles were obtained compared with the samples without MMT. The anti-crack property of the PLA/MMT composite particle may lead to higher encapsulation efficiency because it prevents the breaking leakage of water phase, which contains the hydrophilic drug molecules.

432431 Biocompatibility of Lipid Coated Nanocomposites

Abstract:

There is significant interest in the use of a variety of nanoparticles for drug delivery applications. Solid nanoparticles are unique in that they enable incorporation of drug on the surface of nonporous particles as well as within mesoporous systems. As such, nanoparticles have been examined for use in many routes of delivery such as pulmonary, transdermal, oral and intravenous. Despite the breadth of research examining their use as drug delivery platforms, there is a lack of information regarding the health and safety of solid nanomaterials. In most studies investigating cell toxicity little regard is given to the characterization of the nanomaterials being used. Important aspects often left unconsidered include size, size distribution, surface charge, aspect ratio, morphology, and preparation method.

In addition to traditional solid nanoparticles, liposomes have been highly researched and represent one of the few nanocarriers to successfully transition to clinical applications. Studies of liposome toxicity have shown a strong correlation to their composition. Surface charge, size and concentration have been shown to be significant factors affecting toxicity of liposomes.

Recent literature has focused on combining nanomaterials to develop novel nanocarriers such as liposome encapsulated solid nanoparticles. However, little concern has been given to the health and safety of these newly formed nanocomposites. This study aims to elucidate the missing cellular toxicity effects associated with these new materials and provide a direct comparison to traditional solid nanoparticles. In addition to lipid encapsulation, the factors investigated included size-dependence, dose-dependence, and surface modification effects.

433024 Biomimetic Amphiphilic Polymers for Intracellular Therapeutic Delivery and Theranostic Applications

Abstract:

Biomacromolecules represent a powerful new class of medicine with potential for treatment of a wide variety of previously intractable human diseases. However, it remains a big challenge to effectively deliver them through extracellular and intracellular barriers. There is a need to better understand the mechanisms of entry into the cell cytoplasm and nucleus in order to design optimal delivery systems for biological molecules. On the one hand, this would open up significant opportunities to deliver potent drug payloads against intracellular targets to positively impact human health. In addition this would enable us to develop a more general understanding of the rules governing the uptake of biological molecules into cells. This presentation will cover our recent efforts to design, synthesis and in-vitro/in-vivo evaluation of novel bio-functional polymers. The pH-responsive, metabolite-derived, amphiphilic polymers are designed to mimic factors that enable efficient viral transfection. Strict control over the size, structure, hydrophobicity-hydrophilicity balance and aromaticity of the polymers can effectively manipulate their conformational characteristics and dynamic behaviour in aqueous solution and their interactions with lipid membrane, cell and tissue models. It has been demonstrated that the biomimetic polymers can traverse the extracellular matrix in 3-D multicellular spheroids, reach individual cells in the tumour models, and enable efficient cellular entry and substantial endosomal release of payloads (e.g. siRNA and therapeutic proteins) and imaging moieties into the cytoplasm or the nucleus. This could represent a promising therapeutic delivery platform, suitable for research and theranostic applications in the treatment of various diseases including cancer.

437859 Water Desalination By Shock Electrodialysis

Abstract:

The development of cost and energy efficient water purification systems, specifically desalination systems, is one of the most critical engineering challenges today. Recently, desalination has been used to solve the drought crisis in California, but it is costly and energy intensive [1]. Shock electrodialysis is a newly developed method that uses ion concentration polarization (ICP) zones and deionization shocks in porous media near an ion selective membrane to purify water. In this project, we demonstrate experimentally the feasibility of shock electrodialysis in water desalination. An overlimiting current, a current that is more than the diffusion limited current, is applied through a weakly charged porous medium (a silica glass frit with pore sizes ranging from 0.75 to 1 micron) sandwiched between two cation-selective membranes (Nafion), producing a sharp concentration gradient or "shock"[2]. The cations move through the cation-selective membrane toward the cathode while the anions get trapped on the cation-selective membrane on the anode side. This movement of ions produces a sharp concentration gradient with high ion concentration near the anode side and low ion concentration near the cathode side. Flowing saltwater through the porous medium and adding a splitter that splits the flow into two streams at the end can separate the high ion concentration (brine) stream and the low ion concentration (fresh) stream. This separation allows for continuous desalination. Using this device, we were able to remove up to 99+% of ions from NaCl feed solutions of various concentrations (1 mM, 10mM, and 100mM) at a flowrate of 75 uL/min. Keeping the flow rate constant at 75 uL/min and varying the concentrations revealed that plots of the percentage of ion removal versus the applied current divided by the limiting current for all three concentrations overlapped, agreeing with the theory. By varying the flow rate and keeping the concentration of NaCl constant at 10mM, it was seen that at lower flowrates, there was more ion removal at the same applied current, agreeing with the theory because a lower flowrate allows the solution to spend more time in the device which allows for more ion removal. The 37.5 uL/min and 75 uL/min flowrates achieved 99+% ion removal at the highest applied current (4080 uA) while the 150 uL/min flowrate achieved 90% ion removal. Overall, preliminary results show good desalination factors, reasonable efficiency, and unique separation capabilities. These unique separation capabilities stem from the feed water flowing through the glass silica and the large electric field applied. The small pore size in the glass silica frit is able to separate out large particles and the large electric field has the potential for disinfection; both of these qualities may find applications in compact water treatment and brine concentration.

[1] Roth, S.; California's last resort: Drink the Pacific; Desert Sun; 2015. http://www.desertsun.com/story/news/environment/2015/04/20/californias-last-resort-drink-pacific/26081355/

[2] Mani, A.; Zangle, T.; Santiago, J.G. On the Propagation of Concentration Polarization from Microchannel-Nanochannel Interfaces Part I: Analytical Model and Characteristic Analysis. Langmuir 2009, 25 (6), 3898-3908

437635 Exergy Analysis of a Power Plant in Abu Dhabi (UAE)

Abstract:

The first objective of this research was to conduct an exergy analysis for a power-generation plant in Abu Dhabi (UAE) in order to determine the main source of its irreversibilities. Our results indicated that the combustion chamber was the main factor (70.2%) of the total exergy destruction of the plant. On the other hand, the compressor had the lowest contribution towards exergy destruction (12.4%). In the second part of this investigation, Aspen Hysys with the Soave-Redlich-Kwong (SRK) equation of state were utilized in order to simulate the effects of high temperatures and absolute humidity of ambient air on the exergy efficiency of the different equipment in the power plant. Our data showed that the temperature of ambient air had more negative effects on the performance of the plant than its absolute humidity. The exergy efficiency of the combustor and compressor decreased with temperature while the exergy efficiency of the turbine increased with temperature. Finally, a comparative study on performance of the power plant between design conditions (T=288K, RH=60%) and summer conditions (T=316K, RH=50%) in Abu Dhabi indicated that the power plant lost 4.66 % of its net power output and 4.61 % of its exergy efficiency.

438525 Insights into the Hydrothermal Stability of ZSM-5 Under Relevant Biomass Conversion Reaction Conditions

Abstract:

Zeolite catalysts used for the conversion of carbohydrates to renewable platform chemicals (e.g., glucose to HMF) in the condensed phase are shown to be sensitive to the presence of inorganic salts which alter the zeolite surface chemistry. Although zeolites such as ZSM-5 have been shown to be stable in de-ionized water, the presence of NaCl in the reaction mixture at dilute and saturated levels enhances the hydrolysis of the Si-O-Al bridges and the release of Al3+ species. These dissolved Al3+ species catalyze the conversion of glucose through homogeneous catalytic processes and obscure the apparent reactivity of the zeolite catalyst. Comparative XRD, solution-state NMR, and ICP-MS show that the reactive aluminum species in solution are leeched in a time-limited process from Si-O-Al bonds involved in extra-framework species and/or at defect sites without affecting the zeolite lattice

EFFECTS OF SEASON AND HEATING MODE ON IGNITION AND BURNING BEHAVIOR OF TEN SPECIES OF LIVE FUEL MEASURED IN A FLAT-FLAME BURNER SYSTEM

Samantha A. Smith, Jonathan R. Gallacher, and Thomas H. Fletcher

Abstract

The effects of season and heating mode on ignition and burning behavior of living vegetation were studied in a flat-flame burner system with a radiant panel. The goal was to identify what plant characteristics have the greatest influence on burning behavior and to understand the effects of heating mode on ignition and burning. Experiments were performed for ten species over a two-year period, with 25 runs completed each month for each species. A flat flame burner (1000°C, 10 mol% O_2) and radiant panel (50 kW/m^2) provided the convection and radiation sources, respectively. Time-dependent mass, surface temperature and flame characteristics were measured. The effect of moisture content on ignition time varied with species. Ignition time showed a strong dependence on heating mode, with broadleaf species showing a much stronger response to added radiation than non-broadleaf species. No samples exposed to the radiant panel alone ignited. Heating mode and moisture content affected flame characteristics and the relationship depended upon season and species. The observed relationship between ignition time and moisture content did not agree with the relationship observed in wet wood. Moisture content alone did not adequately describe seasonal changes in ignition behavior. However, by including other factors in the prediction model (e.g. surface area and mass), it is possible to predict live-fuel ignition. This implies that plant function plays a role in the burning behavior of live fuels.

1. Introduction

Ignition of wood and other cellulosic fuels has been studied for over 100 years. Research has largely focused on requirements for ignition, conditions during burning, and predictive modeling techniques (including rate of spread calculations). The ultimate goal in wildland fire research is two-fold: (1) to understand the physical phenomena that occur within wildland fires, and (2) to develop models that can predict wildland fire behavior. This study presents some of the results from a two-year project to study ignition and burning phenomena relative to seasonal changes in plant growth and the mode of heat transfer used for fuel heating.

Ignition can occur when a fuel sample is heated to the point where pyrolysis rates are high enough to support a gaseous flame *and* a gas phase flammable mixture exists. Researchers often measure ignition temperature and time. These values are then used as empirical estimates of the surface temperature at which pyrolysis rates can support a continuous flame and the time required to reach that temperature, respectively. The bulk of the work has focused on ignition temperature. Experimental conclusions to date are mixed. Babrauskas [1, 2] compiled the results of ignition temperature experiments on wood fuels and foliage, respectively. After eliminating the experiments in which the fuel sample was pressed against a hot surface, the reported ignition temperatures ranged from 200-530°C for wood and 201-450°C for foliage. Babrauskas noted the large amount of scatter in the data and suggested that, in addition to variations in experimental setup and measurement techniques, sample condition (e.g. moisture content and size) and species could affect ignition temperature.

Wildland fire observations that species burn differently support Babrauskas's postulate that plant species could be one source of variation in measured ignition temperatures [3]. However, results by Susott [4] showed that material ground from various plant species has the same heat of combustion and similar TGA (thermogravimetric analysis) pyrolysis mass release curves, and should therefore burn similarly. Thus, one possible explanation for the observed differences in ignition properties is the shape and structure of the plant and the effect shape has on heat and mass transfer. However, this explanation has not been tested experimentally. Most empirical correlations used to predict ignition behavior, particularly for live fuels, are species specific [5-8]. Work must be done to understand the differences in ignition behavior between various species.

Investigation of the effect of moisture content on ignition has been studied extensively and supports Babrauskas's postulate that sample condition affects ignition. Most of the work has focused on dead or woody fuels, with the result that moisture increases both ignition time and temperature [5, 6, 9-11]. The results for live fuels are fewer and less consistent [3, 12-15]. Many fire spread models use moisture content as a predictor for ignition. These models have varied success with dead fuels but little success with live fuels [13]. One possible reason for the difference in ignition between live and dead fuels and the performance of fire spread models in predicting fire behavior in live fuels is the difference in water storage between live and dead plants and the resulting treatment of water storage in fire spread models. Water evaporation in dead fuels has been assumed complete in fine fuels once the sample temperature passes 100°C [16, 17], but Fletcher et al. [3] showed there is still a significant amount of moisture in live fuels when ignition occurs. Pickett [18] showed water release still occurring at surface temperatures in excess of 200°C and Prince [14] showed significant differences in the temperature profiles of live and dead foliage during ignition and burning even with the same moisture content. Work by McAllister et al. [13] showed significant differences in the ignition behavior between live and dead pine needles. These differences have led researchers to postulate that there is significant interaction between the free water and the cells in live plants that does not occur in dead plants [13, 19]. Finney et al. [20] postulated that water release in live fuels is not complete until the breakdown of the cellular structure occurs. Still other work has been done indicating root structure [21], plant dry mass [22], and chemical composition [13, 23] could have a larger effect on ignition of live fuels than moisture content, though results are mixed in work to quantify the effect of chemical composition (e.g. [15]). Thus, a fundamental understanding of the physical processes that drive live fuel combustion is both absent and necessary if predictive models are to be developed.

Another of Babrauskas's postulates was that differences in ignition temperature arise due to differences in experimental setup. This is particularly true when looking at the different heat transfer mechanisms. Much of the work looking at heat transfer mode has focused on dead and woody fuels [10, 17, 24-31]. Experiments performed by Rothermel [17] showed fuel pre-heating in no-wind and backing-fire situations, leading to the conclusion that radiation drives fire spread. This conclusion and his resulting fire spread model established a precedent that radiation is the dominant form of heat transfer in wildland fires. However, other experiments have shown that significant amounts of pyrolyzates are not formed at the fuel temperatures associated with radiant pre-heating [32]. Anderson [33] concluded that radiant heat flux can provide no more than 40% of the energy required for sustained fire spread. Still other work has shown flame propagation to depend strongly on direct flame contact with un-burned fuel [34, 35]. Many researchers have concluded that radiation heat transfer dominates in large fires [24, 31, 36, 37] and fires with little to no wind in homogeneous fuel beds [28, 38-40], but the relative effect of radiation and convection for fires outside these conditions is still unknown [28, 40]. Only a limited amount of work has been performed for live fuels and foliage [13, 41]. Work to quantify the contributions of radiation and convection in live-shrub combustion is necessary to understand the basic theory of fire spread and to develop a model that accounts for both mechanisms of heat transfer.

The aim of this project is to explore the effect of season and heating mode on ignition and burning behavior to better understand what physical processes drive fire spread in live shrubs.

2. Methods

Combustion experiments were performed on ten species every month for two years. Each month 25 replicates were performed for each species. The ten species consisted of three species from California, three from Florida, two from Utah, and two from Montana. The plants from California consisted of chamise (*Adenostoma fasciculatum*), Manzanita (*Arctostaphylos glandulosa*), and Ceanothus (*Ceanothus crassifolius*). Gallberry (*Ilex glabra*), fetterbush (*Lyonia lucida*), and sand pine (*Pinus clausa*) originated from Florida, while big sagebrush (*Artemisia tridentata*), and Gambel oak (*Quercus gambelii*) are from Utah. Finally, lodgepole pine (*Pinus contorta*) and Douglas-fir (*Pseudotsuga menziesii* var. *glauca*) are from Montana. Since most of the samples had to be collected from other locations, they were placed in a sealed plastic bag and shipped overnight, where they were placed away from light with the bag still closed until that species could be tested.

Experiments were performed using a water cooled, flat-flame burner (FFB) and radiant panel (see Figure 1). The FFB produced post-combustion gases at 1000 °C and 10 mol% O_2 while the radiant panel provided 50 kW m^{-2} at the sample location. A glass cage prevented ambient air entrainment. In order to perform the experiments, the sample was suspended within the cage using an alligator clip and holding rod. This rod was placed on a Mettler Toledo XS204 Cantilever mass balance, which recorded any differences in mass as the specimen burned or experienced pyrolysis. Each month, experiments were performed using three heating cases, namely convection, radiation, and combined convection plus radiation. In all cases there was a visual camera placed in front of the sample and an infrared camera placed above the sample (see Figure 1).

Transient flame, mass, and surface temperature data were collected during each run. Flame characteristics were extracted from these data, including maximum flame height (cm), time to maximum flame height (s), ignition temperature (°C), maximum ignition temperature (°C), ignition time (s), and burn out time (s). There were also many measurements of the specimen itself, such as the density, surface area, moisture content, length, width, thickness, etc.

Both single and multiple parameter prediction models were developed to describe each of the flame characteristics listed above, namely time to ignition (t_{ig}), surface temperature at ignition (T_{ig}), maximum surface temperature at ignition ($T_{ig,max}$), maximum flame height (MFH), time to maximum flame height (t_{MFH}) and burnout time (t_{BO}). Ignition has been shown to be a local phenomenon [19]. Surface temperature has also been shown to have large spatial variations during the burning of live foliage [19]. However, many ignition temperature measurements do not measure the spatial variation in surface temperature. Correlations for both average surface temperature at ignition and maximum surface temperature at ignition (assuming the location of maximum temperature corresponds to the localized ignition point) were developed. One-parameter models were developed using simple linear regression. The F-statistic, p-value, and confidence interval on the slope were used to determine if the slope term and regression model were significant.

Forward and backward stepwise regression was used to develop correlations with multiple parameters. To do this, 500 correlations were developed by randomly selecting four parameters for use in each model and by randomly selecting a transformation of the data for those parameters. To develop the models, the parameters were first assigned to groups based on type of measurement to reduce the chance of multicollinearity. The groups were: (1) moisture content, relative moisture content and water mass; (2) fresh mass and dry mass; (3) length, width and surface area; (4) needle length, stem diameter, thickness and density; and (5) lipid content, volatile fraction, fixed carbon and ash content. For each correlation, four of the five groups were randomly chosen and then a parameter from each of those four groups was randomly selected. Once the parameters were identified, a data transformation was applied to the data for each parameter. The transformations were: (1) no change, (2) square root of the data, (3) square of the data, (4) natural log of the data, and (5) inverse of the data. Once the data and data transformations were assigned, stepwise regression was performed to develop a correlation to fit the selected data to each of burning characteristics listed above. Once the 500 models were created, the model with the highest adjusted R^2 for each burning characteristic for each species was identified.

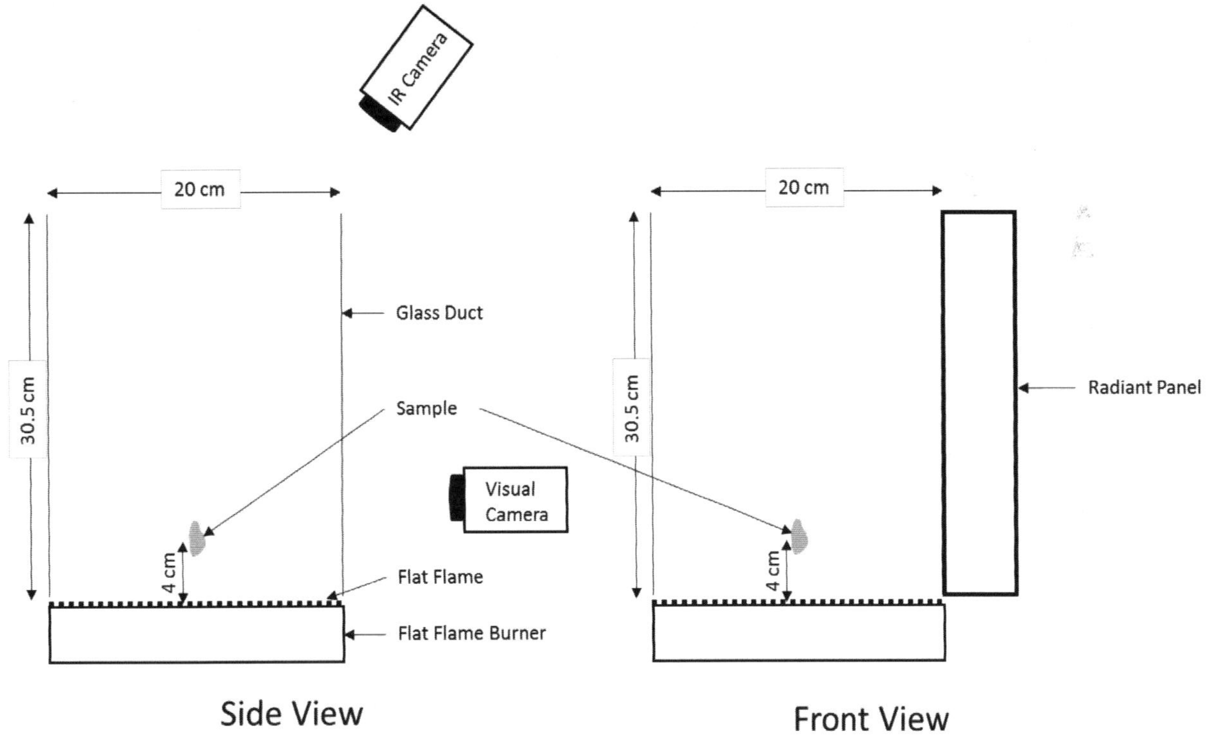

Figure 1. Experimental apparatus: flat flame burner, radiant panel, IR camera, video camera, glass cage to prevent ambient air entrainment, sample location, mass balance and sample holding rod. Post-flame gases are 1000 °C and 10 mol% O_2.

3. Results and Discussion

It was predicted that an increase in the moisture content of a species would result in a longer ignition time and would closely resemble the burning of wet wood. This was not the case. Table 1 illustrates how live fuels responded to different burning modes during the year for different flame characteristics. If there was a statistically significant positive or negative slope at the 95% confidence level, then a P or N was indicated, respectively. Zeros indicate that there was no statistically significant relationship. Overall, the results from the live fuels were different from the expected behavior with wet wood illustrating that one-parameter models do not work for live fuels. Ceanothus, Gambel Oak, Douglas-fir, and Chamise, however, did exhibit a positive correlation to the wet wood behavior at the 95% confidence level.

Table 1. Significance of yearly trends by species

	Ignition Time (s) vs MC		**MFH (cm) vs MC**		**Ignition Devol (%/s) vs MC**		**Ignition Devol (%/s) vs Ignition Time (s)**	
Species	Conv*	Comb**	Conv	Comb	Conv	Comb	Conv	Comb
Manzanita	0	0	0	0	0	0	N	0
Ceanothus	P	P	0	0	0	0	0	0
Douglas-fir	P	0	N	0	0	0	0	0
Gambel Oak	P	P	N	N	0	0	0	0
Fetterbush	0	0	0	0	0	0	N	0
Gallberry	0	0	0	0	0	0	N	0
Sand Pine	0	P	0	0	0	0	0	0
Chamise	P	P	0	0	0	0	0	0
Sagebrush	0	0	0	0	0	0	0	0
Lodgepole Pine	0	0	0	0	0	0	0	0
Wet Wood (expected)	P	P	N	N	P	P	P	P

*Conv means convection-only experiments; **Comb means combined convection and radiation

The models created using stepwise regression were much better than the one-parameter models. Table 2 shows the adjusted R^2 values for the flame characteristics measured with these new models. The average adjusted R^2 value for all the models is 0.406. However, the models for temperature account for much less of the data variation than the models for the other flame characteristics; the average adjusted R^2 value for the non-temperature models in 0.569. This is partly due to the fact that samples bend and move during burning and thus the entire sample surface was not visible to the IR camera during the run.

Table 2: Adjusted R^2 values when regressing flame characteristics using stepwise models.

Species	t_{ig}	T_{ig}	$T_{ig,max}$	MFH	t_{MFH}	t_{BO}
Manzanita	0.80	0.527	0.568	0.740	0.919	0.747
Ceanothus	0.794	0.481	0.372	0.743	0.905	0.676
Douglas-fir	0.303	0.571	0.692	0.342	0.254	0.50
Gambel Oak	0.664	0.472	0.186	0.432	0.86	0.81
Fetterbush	0.537	0.379	0.058	0.431	0.417	0.459
Gallberry	0.733	0.152	0.223	0.683	0.743	0.511
Sand Pine	0.434	0.294	0.247	0.58	0.624	0.639
Chamise	0.554	0.286	0.305	0.617	0.502	0.449
Sagebrush	0.449	0.285	0.247	0.403	0.589	0.403
Lodgepole Pine	0.489	0.315	0.309	0.384	0.36	0.30

Figure 2 illustrates how ignition time and moisture content vary throughout the year for Douglas-fir and Manzanita with different heating modes (convection-only, and combined convection and radiation). It is observed that there is no statistically significant relationship between moisture content and ignition time. Manzanita had approximately the same moisture content in July and November, but the ignition times in July were about half as long as the ignition times in November. Douglas-fir moisture content doubled from May to September, yet there was no significant change in the ignition time. These examples point toward the initial observation that the single parameter model of just moisture content cannot provide an accurate prediction for live fuels (see Table 1). Since the moisture content isn't the only thing that is impacting the burning of the live fuels, it is logical to consider the plant's physiological structure and how that can change the way it burns. The structure of the plant does change throughout the seasons and that can change how the species stores the water and how the leaves are shaped. These changes can help account for the changes that are not accounted for by just the water in the plant. It is important to note that, while a study of how physiological factors impact the burning of the live fuels is needed to develop a deeper theoretical understanding of live fuel combustion, it is still possible to predict burning behavior with statistical models that use the same inputs as current fire spread models.

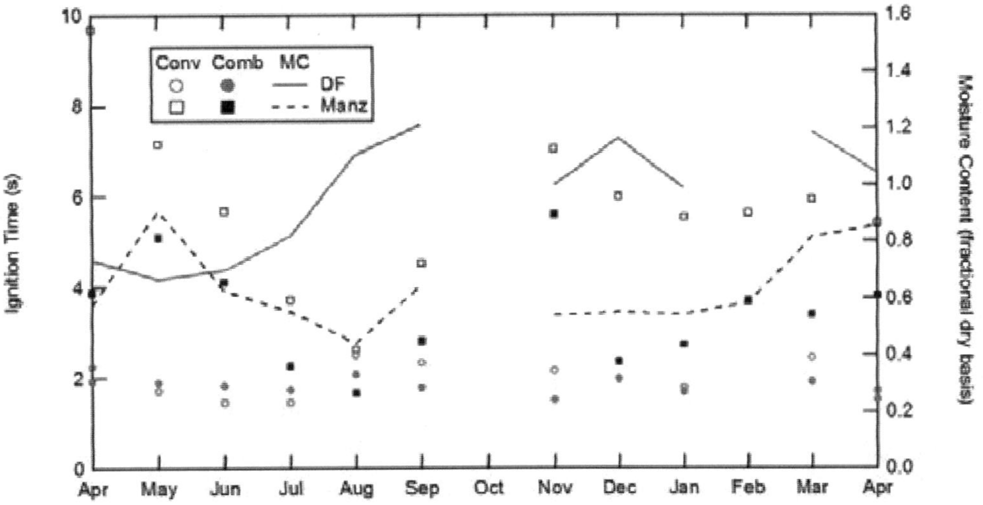

Figure 2. Ignition time and moisture content versus month for manzanita and Douglas-fir. Closed symbols indicate radiation and convection ignition while open symbols indicate convection-only ignition.

In addition to the models that were produced, there was data collected on the different heating modes and how they impact the burning of the live fuels. Table 3 shows the percentage of months there was a statistically significant difference between the convection-only and convection and radiation runs for each species. There is a noticeable difference between the broadleaf species and the needle species for the ignition time and time to maximum flame height due to the added radiation. Radiation alone was never enough to ignite fuel samples without a pilot ignition source. The heat did, however, cause the samples to undergo pyrolysis and then char oxidation.

Table 3. Effect of heating mode on ignition variables. Table entries indicate the percentage of months that radiation and convection ignition differed significantly convection-only ignition at a 95% confidence level

Species*	Ignition Time	Time to Max Flame Height	Maximum Flame Height	Fraction Left at Ignition	Fraction Loss at Ignition
Manzanita	92	83	83	25	42
Ceanothus	100	91	100	45	18
Gambel oak	83	100	83	50	17
Fetterbush	100	100	8	17	25
Gallberry	92	100	0	17	58
Dougals-fir	9	27	27	0	9
Sand pine	25	25	8	0	17
Chamise	33	25	17	17	0
Sagebrush	22	22	22	11	11
Lodgepole pine	50	25	13	38	13

*The first five species listed are broadleaf species, the other five are non-broadleaf.

4. Summary and Conclusions

Ignition and burning behavior for ten live fuels were studied in a flat-flame burner apparatus. Experiments were performed over a two-year period to test the effect of season (specifically moisture content) on ignition and burning behavior. The effect of heat transfer mode was also tested by performing some of the seasonal experiments with a convection-only heat source, some with a radiation-only heat source, and some with both heat sources.

Results comparing ignition and burning characteristics with moisture content and season were mixed. Ceanothus, Gambel oak, Douglas-fir and chamise all exhibited a positive correlation between ignition time and moisture content while the other six species show no correlation at a 95% confidence level, indicating a simple relationship between moisture content and ignition is not adequate to describe ignition in live fuels. However, linear stepwise models capture much of the variability in ignition behavior. The ignition behavior of live fuels in different seasons but at the same moisture content was different. These results suggest a significant difference between live fuels and dead fuels that isn't explained by moisture content. An in-depth study involving careful combustion experiments and detailed physiological measurements is necessary to develop a theoretical understanding of fire spread in live fuels. In the absence of a theoretical understanding, simple statistical models can still be developed that describe fire behavior accurately and that use as inputs the same information currently used in most fire models.

Heating mode results were clearer. Ignition occurred in none of the unpiloted radiation-only experiments. These results do not indicate radiation was unimportant, but rather that additional modes of heat transfer were needed. Inclusion of a radiant flux in the convective environment of a flat-flame burner significantly decreased observed ignition times for broadleaf species but not for non-broadleaf species. Maximum flame height time showed the same dependence as ignition time for all species while maximum flame height showed the opposite dependence, except for fetterbush and gallberry which showed almost no difference between convection-only and radiation with convection experimental runs. Mass loss rate and fraction remaining at ignition were affected much less by the differences in heating than other ignition characteristics.

5. References

[1] Babrauskas, V., J. of Fire Protection Eng., 12 (2002) 163-189.

[2] Babrauskas, V., Ignition handbook, Fire Science Publishers, 2003.

[3] Fletcher, T. H., Pickett, B. M., Smith, S. G., Spittle, G. S., Woodhouse, M. M., Haake, E.Weise, D. R., Comb. Sci. and Tech., 179 (2007) 1183-1203.

[4] Susott, R. A., For. Sci., 28 (1982) 404-420.

[5] Xanthopoulos, G.Wakimoto, R. H., Canadian J. of Forest Research, 23 (1993).

[6] Dimitrakopoulos, A. P.Papaioannou, K. K., Fire Tech., 37 (2001) 143-152.

[7] Smith, S. G., Effects of moisture on combustion characteristics of live California chaparral and Utah foliage, Brigham Young University (2005).

[8] Shen, C., Application of fuel element combustion properties to a semi-empirical flame propagation model for live wildland Utah shrubs, Brigham Young University (2013).

[9] Gill, A. M.Moore, P. H. R., Ignitability of leaves of Australian plants, A Contract Report to the Australian Flora Foundation, Centre for Plant Biodiversity Research, 1996.

[10] Moghtaderi, B., Novozhilov, V., Fletcher, D. F.Kent, J. H., Fire Safety J., 29 (1997) 41-59.

[11] Catchpole, W. R., Catchpole, E. A., Tate, A. G., Butler, B. W.Rothermel, R. C., A model for the steady spread of fire through a homogeneous fuel bed, 2002.

[12] Engstrom, J. D., Butler, J. K., Smith, S. G., Baxter, L. L., Fletcher, T. H.Weise, D. R., Comb. Sci. and Tech., 176 (2004) 1577-1591.

[13] McAllister, S., Grenfell, I., Hadlow, A., Jolly, W. M., Finney, M.Cohen, J., Fire Safety J., 51 (2012) 133-142.

[14] Prince, D., Measurement and modeling of fire behavior in leaves and sparse shrubs, Brigham Young University (2014).

[15] Alessio, G. A., Penuelas, J., De Lillis, M.Llusia, J., Plant Biology, 10 (2008) 123-128.

[16] Albini, F. A., Proc. Comb. Inst., 11 (1967) 553-560.

[17] Rothermel, R. C., A mathematical model for predicting fire spread in wildland fuels, Research Paper INT-115, USDA Forest Service, 1972.

[18] Pickett, B. M., Effects of moisture on combustion of live wildland forest fuels, Brigham Young University (2008).

[19] Prince, D.Fletcher, T. H., A combined experimental and theoretical study of the combustion of live vs. dead leaves, Combustion Institute - Western States Section (WSS/CI), 2013.
[20] Finney, M. A., Cohen, J. D., McAllister, S. S.Jolly, W. M., Int. J. of Wildland Fire, 22 (2013) 25.
[21] Pellizzaro, G., Duce, P., Ventura, A.Zara, P., Int. J. of Wildland Fire, 16 (2007) 633-641.
[22] Jolly, W. M., Hadlow, A. M.Huguet, K., Int. J. of Wildland Fire, 23 (2014) 480-489.
[23] Pyne, S. J., Andrews, P. L.Laven, R. D., Introduction to Wildland Fire, 2nd ed, Wiley, 1996.
[24] Simms, L., Comb. Flame, 4 (1960) 293-300.
[25] McCarter, R. J.Broido, A., Pyro., 2 (1965) 65-85.
[26] Simms, L.Law, M., Comb. Flame, 11 (1967) 377-388.
[27] Pagni, J., J. of Heat Trans., 97 (1975) 153-155.
[28] Morandini, F., Santoni, P. A.Balbi, J. H., Fire Safety J., 36 (2001) 519-543.
[29] Dupuy, J. L., Maréchal, J.Morvan, D., Comb. Flame, 135 (2003) 65-76.
[30] Reszka, P.Torero, J. L., Exp. Therm. and Fluid Sci., 32 (2008) 1405-1411.
[31] Silvani, X.Morandini, F., Fire Safety J., 44 (2009) 279-285.
[32] Cohen, J. D.Finney, M. A., VI International Conference on Forest Fire Research, (2010).
[33] Anderson, H. E., Heat transfer and fire spread, in Research Paper INT-69. 1969, USDA Forest Service.
[34] Vogel, M.Williams, F. R., Comb. Sci. and Tech., 1 (1970) 429-436.
[35] Carrier, G. F., Fendell, F. E.Wolff, M. F., Comb. Sci. and Tech., 75 (1991) 31-51.
[36] Balbi, J.-H., Rossi, J.-L., Marcelli, T.Santoni, P.-A., Comb. Sci. and Tech., 179 (2007) 2511-2537.
[37] Paudel, A., Application of one dimensional turbulence (odt) to model fire spread through biomass fuel bed, Brigham Young University (2013).
[38] Morvan, D.Dupuy, J. L., Comb. Flame, 127 (2001) 1981-1994.
[39] Morvan, D.Dupuy, J. L., Comb. Flame, 138 (2004) 199-210.
[40] Sullivan, A. L., Ellis, P. F.Knight, I. K., Int. J. of Wildland Fire, 12 (2003) 101-110.
[41] Stocks, B. J., Alexander, M. E., Wotton, B. M., Stefner, C. N., Flannigan, M. D., Taylor, S. W., Lavoie, N., Mason, J. A., Hartley, G. R., Maffey, M. E., Dalrymple, G. N., Blake, T. W., Cruz, M. G.Lanoville, R. A., Canadian J. of Forest Research, 34 (2004) 1548-1560.
[42] ASTMStandardD4442-07, Test methods for direct moisture content of wood and wood-based materials, in ASTM International. 2007: West Conshohocken, PA.
[43] Weber, R. O., Prog. Energy Combust. Sci., 17 (1991) 67-82.
[44] Sullivan, A. L., Int. J. of Wildland Fire, 18 (2009) 349-368.

438326 Characterization and Particle Sizing of the Composition of E-Cigarette Aerosol

Abstract:

In recent years, electronic cigarettes, or e-cigarettes, have rapidly gained immense popularity throughout the world and specifically the United States. Despite the increase in use, little is known about the health effects of these vaporizers, compared to traditional cigarettes. The e-cigarette uses a wick and heated coil to atomize e-liquid, creating the aerosol inhaled by the user. E-liquids are composed of propylene glycol, glycerin, usually nicotine, and various flavoring agents. E-cigarettes were customized in order to simulate a puff on the device by pushing a known volume of air through. An Automated Puffer Machine (APM) was designed to control power delivered to the coil, puff volume, puff frequency, and puff duration. An Electrostatic Classifier (EC) and Condensation Particle Counter (CPC) then sampled a diluted puff, analyzing particle size and concentration. A Poly Urethane Foam (PUF) sampler was used to separate and trap the chemical components of the e-cigarette emission. The PUF was then placed in a volatile solvent to dissolve the particulate and vapor components, and the solvent was analyzed using gas chromatography with mass spectroscopy. The PUF sample was used because is closely emulates what a user would normally inhale.

Work has also been done to quantify and describe the aerosol released by electronic cigarettes in order to create profiles of the particles released into the atmosphere. Examples include work to quantify particle size distribution and chemical content of the aerosol in large gas chambers and looking at particle size and distribution with rapid sampling. Particle size of the aerosol was a bimodal distribution, with main peak averages of 30 nanometers and 80 nanometers. Research isolated the components of the e-liquid attempting to sort which vaporized component contributed to the collected particle size profile. Further research attempted to determine the amount of nicotine available within both the vapor and particle components of the aerosol and identify the exact composition of the aerosol created by an e-cigarette when the e-liquid is vaporized.

This study has found that a variety of e-liquids ranging in nicotine contents and flavors consistently produce a bimodal distribution of aerosol particle sizes. Furthermore, this study concludes that the proportion of components effects the concentration of particles of the aerosol.

438306 Ammonia Removal from Aquaculture Stocking Water

Abstract:

A continuous membrane system capable of maintaining low levels of total ammonia nitrogen (TAN, 3, and 4 +) in aquaculture holding and transport vessels is being developed. Ammonia is the major metabolite of fish, therefore, adequate TAN removal is necessary to maintain the health and productivity of aquatic organisms. High levels of ammonia result in the damage of gas exchange tissues, leading to increased mortality and decreased average harvest weight. Membrane systems are desirable for their scalability, lack of compaction and channeling, as in packed beds, and their ability to stabilize additives used during functionalization. Zeolite 13X demonstrates efficient 4 + exchange with + in aqueous solutions. Zeolite 13X is inexpensive and functions well as an ion exchange material with high ammonium exchange capacities. In order to improve the stability of zeolite 13X in water and develop a continuous TAN removal scheme, mixed-matrix membranes have been fabricated in the presence and absence of porogens. Porogens have been used to investigate the effect of pore size and pore density on the membrane structure and the membrane performance. The membranes with porogens exhibited higher flux but lower ammonia removal due to decreased residence time. Porogen-free membranes had a higher ammonia removal capacity due to longer residence time. The ammonia removal capacity was found to correlate well with the amount of zeolite 13X incorporated. An optimal formulation will maximize flux and available zeolite surface area, minimize residence time, and retain large ammonia loadings after a large volume of liquid has been treated.

438394 Impact of Chemical Dopants and Passivation Schemes on Carbon Nanotube Sheet Conductivity

Abstract:

Carbon nanotubes (CNTs) are being investigated as an alternative to conventional metallic conductors. Improving the electrical performance of bulk CNT materials through the use of chemical dopants is an active area of research. The present work focuses on doping efficacy and dopant stability under various environmental conditions. CNT sheet material was doped with a series of compounds of varying structure, and a 3 to 8-fold increase in electrical conductivity was observed after doping with potassium tetrabromoaurate, iodine, or iodine monobromide. The general trend for the doped samples in ambient conditions was a steady decrease in electrical conductivity over several months. These experiments served as an initial assessment of doped CNT sheet stability for conventional conductor applications and demonstrate the necessity towards passivation schemes for this emerging technology. Different encapsulation strategies are currently being explored to retain the conductivity of the films when doped. The conductivity of these samples is being systematically measured while storing them under ambient conditions. Consideration for the encapsulant type, deposition technique, and solvent effects (which may dissolve the dopants from the CNT network) will be characterized. A summary detailing the progress to date will provide an overview of both the best methods to enhance electrical conductivity in CNT sheets as well as the suitable passivation schemes for practical operation.

438530 Îœfilms: Dynamics of Thin-Films Under Physiological Fluids and Shear Flow

Abstract:

Thin-film drug-delivery is an effective alternative to conventional tablets, capsules and liquids as it improves medicinal bioavailability, efficacy, dosage accuracy and onset of effectiveness. Typically, it consists of a thin, dissolving film that orally administers drugs. The significantly large surface area of the thin-film causes rapid disintegration, releasing drugs into systemic circulation via absorption through the oral mucosa. Thin-films are currently studied under static conditions that do not account for the influence of dynamic variables like shear stress due to blood flow, changes in electric potential and interactions with biological fluids on drug delivery. Here we seek to study films under dynamic physiological conditions that closely mimic those within the body, using a microfluidic chip called μFILMS. In the talk, we show our initial results, chiefly construction of the μFILMS chip using additive manufacturing, analysis of the output signals and biochemical/mechanical adjustment of the chip parameters to optimize the visualization and/or characterization of the films. The construction of the μFILMS chip consists of a Transwell® static membrane placed between two glass slides, forming two distinct

438444 Photocatalytic Methanol Reforming on TiO2

Abstract:

Hydrogen gas has great potential for use in transportation fuels and industrial applications, but as it is currently produced commercially from steam reforming of unrenewable natural gas. An alternative source of hydrogen could be accessed from the reforming of sustainably-grown biomass. Through the use of a photocatalyst, biomass can be decomposed, producing clean-burning fuel. Titanium dioxide (TiO_2) is a metal oxide semiconductor of particular interest in photocatalysis due to its activity under ultraviolet illumination and its stability under catalytic reaction conditions. Ordinary TiO_2 (anatase, rutile) only absorbs UV wavelengths. Since a much larger fraction of solar irradiation is in the visible range, we aim to modify TiO_2-based photocatalyst to interact with visible light. Supporting metal nanoparticles on the surface of TiO_2 is known to cause a red shift in the absorbance of TiO_2. We have supported copper, nickel, silver, platinum, and palladium on TiO_2 through insipient wetness impregnation. The metal-modified TiO_2 was tested in two types of in situ reactors; one inside a simple kinetic reactor, and the other in a UV-Visible Spectrophotometer. The samples were then reacted under UV and visible light. Gaseous mixtures of methanol, oxygen and steam were passed over the catalysts while the temperature of the reaction was externally controlled. The products of the reaction were identified by gas chromatography. The addition of each of the metals shifted the absorbance of the TiO_2 to longer wavelengths. Future work includes further characterization of the materials, as well as varying the particle size of the metals and the support, and doping the support with hydrogen, carbon, and nitrogen.

438531 Room Temperature Shape Memory Polymers

Abstract:

Temperature controlled shape memory polymers can be synthesized via a facile method using long chain fatty acid and cross-linked natural rubber. For this experiment, rubber was swollen in liquid stearic acid (75 °C). Different weight percent loadings of acid were achieved by differing swelling times in one hour increments from 1 hour to 5 hours. The normal shape memory cycle consist of increasing the temperature of the swollen polymer above the melting point of the acid, stressing the polymer, then immediately quenching the heat in room temperature water. The rubber will hold stress without relaxing until heated above the melting point of stearic acid again, thus returning the rubber to its original state. Obtaining shape memory behavior at temperatures lower than 75 °C is desirable to mitigate safety concerns and reduce heating required, improving process economics. Investigation of the room temperature shape memory behavior was conducted through modifying the shape memory cycle. It was found that stearic acid swelled natural rubber exhibits a 30% fixity over a 24 hour period. The stress can be relaxed by heating the swelled rubber above the melting point of stearic acid.

429688 Suppression of Infrared Absorption in Nanostructured Metals By Controlling Faraday Inductance and Electron Path Length

Abstract:

While metals are widely used for electrodes in many optoelectronic devices, the device efficiency is limited by optical losses due to metals. As thinner devices are desired for reduced cost without sacrificing performance, the metal electrodes can be nanostructured to concentrate light in the photoactive material. A popular approach for the light concentration is to excite surface plasmon polaritons (SPPs). However, in such SPP-based devices, strong metal absorption is a major source of losses. The problem of this parasitic absorption becomes critical when a high device efficiency is desired. Moreover, if the device operates over a broad band as in solar cells, metal absorption needs to be controlled over the whole spectrum of interest and this poses a significant scientific and engineering challenge. Here, we investigate the loss suppression in metal nanocoil structures in the infrared range. Surprisingly, we find that the nanostructured metals can behave almost like a vacuum with a negligible optical loss in the infrared region. The loss is reduced in comparison to flat films by more than an order of magnitude over most of the very broad spectrum between short and long wavelength infrared. Our detailed analysis shows that this suppression of optical losses is due to both a large Faraday inductance and a long electron path. These two factors render the conduction electron in the metal effectively heavier than a hydrogen molecule. To investigate the usefulness of this effect in infrared detectors, we include a photoactive material in the coiled metal structures. In this case, the fraction of absorption in the active material increases by two orders of magnitude compared to non-coiled structures. Moreover, we demonstrate that the coiling of the infrared detector nanostrips increases the photoresponsivity by 15 times without additional electrical losses. These findings could benefit many metal-based applications that require low loss such as photovoltaics, photoconductive detectors, solar selective surfaces, infrared-transparent defrosting windows, and other metamaterials.

415043 Plasmon-Enhanced Energy Transfer and Other Photophysical Effects in Doped-Lanthanide Nanocrystals

Abstract:

Upconversion of infrared radiation into visible light has been investigated for applications in biological imaging and photovoltaics. However, low conversion efficiency due to small absorption cross-section for infrared light (Yb^{3+}), and slow rate of energy transfer (to Er^{3+} states) has prevented application of upconversion photoluminescence (UPL) for diffuse sunlight or imaging tissue samples. Here, we utilize resonant surface plasmon polaritons (SPP) waves to enhance UPL in doped-lanthanide nanocrystals. Our analysis indicates that SPP waves not only enhance the electromagnetic field, and hence weak Purcell effect, but also increases the rate of resonant energy transfer from Yb^{3+} to Er^{3+} ions by 6 fold. While we do observe strong metal mediated quenching (14 fold) of green fluorescence on flat metal surfaces, the nanostructured metal is resonant in the infrared, and hence enhances the nanocrystal UPL. This strong columbic effect on energy transfer can have important implications for other fluorescent and excitonic systems too. We also investigated the effect of surface plasmons on color tuning and other optoelectronic properties in upconverting nanoparticles. SPP strongly affected the emission color in different-codoped lanthanide UCNPs by tuning the energy transfer between lanthanide ions (Er, Yb and Tm here), depending on the energy overlap between resonant SPP modes and multiple energy states in respective ions. This selective enhancement of resonant energy transfer provides valuable physical insights and useful design principles for electronically coupling different upconverting nanoparticles with matched semiconductor devices and plasmonic absorbers. To test the role of upconversion enhancement in optoelectronics, upconversion enhanced photocurrent was measured on 2D semiconductor nanosheets coupled with UCNPs. The semiconductor 2D nanosheets were nominally transparent to infrared radiation, and no photocurrent is observed on illumination of sub-bandgap radiation. However, strongly enhanced photocurrent was observed in these 2D nanosheets using 980 nm light irradiation, which can enable new applications in development of ultrathin room-temperature infrared detectors and high-resolution imaging systems.

Quantum Dynamical Simulations of the Photoinduced Charge Transfer Process in Donor-Bridge-Acceptor

Ma. Belen Oviedo and Brya M. Wong

1. Department of Chemical Engineering, University of California, Riverside, Riverside, California, United States
2. Materials Science & Engineering Program, University of California, Riverside, Riverside, California, United States

One of the major challenges in science and technology is the efficient conversion of solar energy into readily usable forms. In particular, the enormous size and time scales in solar energy harvesting requires the convergence of multiple scientific disciplines to both improve and optimize these photoconversion processes [1]. Quantum-based simulations play a crucial role in this area since they provide accurate predictions of optical properties of new complex materials. Although very accurate methods exist for predicting optical absorption spectra such as linear-response time-dependent density functional theory [2] or other wave-function based methods, these methods are computationally expensive and can only be applied to relatively small molecules. Consequently, large systems relevant to realistic experiments (such as molecular self-assembled nanostructures and nanomaterials) pose a serious challenge since they lie outside the computational capabilities of these conventional methods.

To this end, we have developed a new real-time time-dependent density functional tight binding (RT-TDDFTB) [3] approach to calculate the electron dynamics of donor-bridge-acceptor complexes. Our approach significantly differs from previous linear-response TD-DFT methods in that we directly propagate the one-electron density matrix in the presence of a non-perturbative external field. Furthermore, this RT-TDDFTB approach permits the calculation of large systems composed of thousands of atoms, enabling new quantum simulations that are beyond the reach of traditional Kohn-Sham approaches in conventional TDDFT. We have successfully used this method to described the photoinduced charge-transfer processes in light-harvesting systems within a fully quantum dynamic framework. In this work Furthermore, we show the charge-transfer state can be modulated by the molecular configuration of the bridge subunit and the charge-transfer processe is enhanced and controlled by the molecular architecture of this subunit.

[1] Nozik, A. J., Miller, J. Chem. Rev., 110, 6443-5 (2010).

[2] Petersilka, M., Gossmann, U. J., Gross, E. K. U. Phys. Rev. Lett., 76, 1212-1215 (1996).

[3] Oviedo, M. B., Zarate, X., Negre, C. F. A., Schott, E., et al. J. Phys. Chem. Lett. 3, 2548-2555 (2012).

415019 Multiple Energy "Exciton-Shelves" in Quantum-Dot-DNA Nanobioelectronic Materials

Abstract:

Quantum dots (QDs) are semiconductor nanocrystallites with size-dependent multiple quantum-confined states which are being explored for utilizing broadband radiation. While DNA has been used for the self-assembly of nanocrystals, it has not been investigated for the formation of simultaneous conduction pathways for transporting multiple energy charges or excitons. These exciton shelves can be formed by coupling the conduction band, valence band, and hot-carriers states in QDs with different HOMO-LUMO levels of DNA nucleobases, resulting from varying degrees of conjugation in the nucleobases. Here, we present studies on the electronic density of states (DOS) in naturally occurring nucleobases (guanine, thymine, cytosine, and adenine), which energetically couple with quantized states in semiconductor QDs. Using scanning tunneling spectroscopy of single nanoparticle-DNA constructs, we demonstrate composite DOS of chemically-coupled DNA oligonucleotides and cadmium chalcogenide QDs (CdS, CdSe, CdTe). While perfectly aligned QD-DNA states lead to exciton-shelves for multiple energy charge transport and energy transfer, mismatched energy levels in QD-DNA hybrids introduce intra-bandgap states which can lead to charge trapping and recombination. Using single nanoparticle and ensemble device measurements we also show successful extraction and transport of both bandedge and high-energy charge carriers, and energy transport using excitons. These results can have important implications for the development of a new class of nano-bio electronics and biological transducers.

419731 Integration of Photosystem I Proteins within Conductive Polymer Matrices Using Vapor Phase Techniques

Abstract:

Organic Solvent-Free Integration of Photosystem I Proteins within Conductive Polymer Matrices Using Vapor Phase Techniques

Maxwell Robinson, Evan Gizzie, David Cliffel, and G. Kane Jennings

Photosystem I (PSI) is a 500 kDa photocatalytic protein complex residing within chloroplast granum of photosynthetic organisms, converting incident solar radiation to electrons with near-unity internal quantum efficiency. Extracted PSI has been employed successfully as a molecular photosensitizer within a host of low-cost electrochemical and solid-state photovoltaic architectures. This presentation explores this group's recent efforts to assimilate PSI within a chemically and energetically tuned conductive polymer matrix. Composite assembly follows a vapor phase Friedel-Crafts grafting procedure, whereby dispersed Fe(III) catalyzes rapid polymer growth from aromatic moieties inherent to PSI. The technique avoids temperature and solvent exposure requirements standard to bulk heterojunction fabrication in organic photovoltaics and can be employed using a wide library of high vapor pressure conjugated monomers. Our composite assemblies enhance charge shuttling processes from individual proteins within multilayer assemblies—greatly reducing charge transfer resistances and improving overall efficiency of photocells.

409549 Solar Evaporation Enhancment Using Floating Copper Oxide Deposited Cellulose Paper

Abstract:

Steam generation using sunlight is usually done by heating the whole body of the water to boiling or higher temprature which leads to lots of heat losses. Recently using nanofluids has been proposed to increase the absorption of solar heat but dispersion of nanoparticles in the water introduce a new set of problems. Here we demonstrtate new strategy for enhancing solar evaporation using floating by depositing the copper oxide nanoparticles on the cellulose paper. CuO nanoparticles were deposited on the paper by a simple in situ method. The resultant nanocomposite showes very good photothermal properties which can be used to enhance the efficiency of solar evaporators. The composite is kept floating on the surface of water by the help of a layer of polysulfone foam underneath it. Preliminary results showes the composite can achieve a solar thermal efficiency of up to 80% with 7 kW/m^2 irradiation. this work provide a low cost and effective way to accelerate solar evaporation for broad range of applications.

433683 Bottom-up Assembly of Metal Silicide Nanowires into Highly Efficient Bulk Thermoelectrics

Abstract:

Increasing the energy efficiencies of the existing processes and systems through waste heat scavenging could be employed to partially offset the projected increase in the future energy demand. The importance of waste heat recovery for energy generation becomes more apparent, when automobiles are considered as an example. In automobiles, approximately 70% of energy generated by burning gasoline is lost either in the engine coolant or through the exhaust, and waste heat scavenging vastly improves their fuel efficiency. Use of solid-state thermoelectric modules is a minimally invasive method for scavenging waste heat from such systems and processes. The lack of any moving parts, portability, very long operational lifetimes, and minimal maintenance make thermoelectrics attractive for waste heat recovery.

The lack of the requisite materials for achieving a good $/watt heat-to-electricity conversion metric is currently preventing the large-scale deployment and use of thermoelectrics in terrestrial applications. A factor responsible for this bottleneck is the low efficiencies of the current thermoelectric materials, which in turn could be attributed to the lack of a pathway for precisely and independently tuning the thermal and electrical transport through them. Recent theoretical and experimental studies have demonstrated that nanostructuring of materials serves as a route for independently tuning the thermal and electrical transport. Most of these demonstrations were made using individual suspended nanowire devices as testpads. However, these studies do not delineate a pathway for extending enhanced thermoelectric performance exhibited by individual nanowires to bulk devices composed of a multitude of nanowires. Metal silicide nanowires are a class of inexpensive and non-toxic materials useful for the fabrication of thermoelectrics, provided their efficiencies are improved beyond that is currently possible. Considering all these facts together, a fundamental question that needs to be answered is the following: How can metal silicide nanowires be mass produced and assembled into devices in a manner that offers precise control over the chemical composition, thermal and electrical transport of the interfaces between the nanowires after their assembly? In this context, we have developed a solid-state phase transformation strategy useful not only for converting pre-synthesized silicon nanowires into metal silicide nanowires, but also for assembling the formed metal silicide nanowires via welding. Assembly of metal silicide nanowires in a manner that makes the chemical composition of the bridges between the nanowires the same as that of the nanowires was accomplished. These results will be illustrated using magnesium silicide (Mg_2Si) nanowire synthesis and assembly as an example. The thermoelectric performances of welded Mg_2Si nanowire assemblies and their relationship to the dimensions of the nanowires and the bridges between the nanowires will be discussed. The thermoelectric performances of welded Mg_2Si nanowire assemblies will be compared to their unwelded counterparts.

426164 Omni-Thermoelectrics: Atomically Convertible p/n Nanowire Inks for Flexible Generators

Abstract:

Thermoelectric devices possess enormous potential to reshape the global energy landscape by converting waste heat into electricity, yet their commercial implementation has been limited by their high cost to output power ratio. No single "champion" thermoelectric material exists due to a broad range of material-dependent thermal and electrical property optimization challenges. While the advent of nanostructuring provided a general design paradigm for reducing material thermal conductivities, there exists no analogous strategy for homogeneous, precise doping of materials. Here, we demonstrate a nanoscale interface engineering approach that harnesses the large chemically accessible surface areas of nanomaterials to yield massive, finely-controlled, and stable changes in the Seebeck coefficient, switching a prototypical *p*-type thermoelectric material, tellurium, into a robust *n*-type material exhibiting stable properties over months of testing. These remodeled, *n*-type nanowires display power factors comparable to their *p*-type counterparts, and are partnered together to demonstrate the first solution-processed, monomaterial flexible thermoelectric generators. We proceed further to dope these nanowires with small organic molecules to generate hybrid organic inorganic nanocomposites and demonstrate power factors and ZTs surpassing bulk tellurium.

432960 Computational Study of Thermal Transport in Si-Ge Nanostructures - Exploration of Phonon Scattering Contributions to Suppressed Conductivity

Abstract:

The rapid emergence of nanostructured materials, which tend to have thermal properties vastly different from their bulk counterparts, has stimulated their utilization for heat dissipation and thermoelectric applications. However, the underlying physical mechanisms that dictate heat conduction still remain unclear. In fact, phonon transport can be impeded by a plethora of possible phonon scattering mechanisms, including phonon-phonon, phonon-boundary, and phonon-impurity scattering, which are difficult to individually characterize experimentally. To investigate these influences, our strategy is to use classical molecular dynamics simulations to explore the thermal transport in Si-Ge nanostructures. In this talk, we present thermal conductivity results for nanostructures including Si-Ge polycrystalline alloys and nanowires. We will discuss the physical origins of the thermal conductivities of these materials by means of the relative contributions of relevant phonon scattering events. We anticipate that this fundamental understanding of phonon scattering behavior will contribute toward the design of nanostructured materials with improved thermoelectric performance.

433904 A Comparative Study of Two Narrow Gap Semiconductors FeGa3 and FeSb2

Abstract:

Introduction

Fe-based semiconductors $FeSb_2$ and $FeGa_3$ have attracted great attention in recent years due to their interesting physical properties. Both of them have fairly small band gaps opened up by the hybridization of the transition metal d orbitals with the main group metal p orbitals. However, the intrinsic energy gap of $FeGa_3$ found above room temperature is one order of magnitude larger than intrinsic energy gap of $FeSb_2$.[1,2] In 2007, Bentien et al. reported a colossal Seebeck coefficient of -45000 μV/K at 10K in single crystal $FeSb_2$. [3] Recently, Seebeck coefficients of a similar order of magnitude (in the order of -16000 μV/K) was also found in $FeGa_3$ single crystalline samples.[4] These two narrow-gap semiconductors therefore are potential candidates for low-temperature thermoelectric application. We are working on the optimization of the crystal growth of these compounds.

Results and discussion

$FeSb_2$ and $FeGa_3$were prepared by self-flux method from mixtures of the pure elements. The structures were refined from single-crystal X-ray data, showing $FeSb_2$ compound crystallizing in orthorhombic space group *Pnnm* (#58) with Z=2 and lattice constants a=5.8328 Å , b=6.5376 Å and c=3.1973 Å, and $FeGa_3$ compound crystallizing in tetragonal space group $P4_2/mnm$ (#136) with Z=4 and lattice constants a=6.2615 Å and c=6.5562 Å.
The resistivity of $FeSb_2$ decreases with increasing temperature which is clearly semiconducting, and a large decrease in resistivity of more than six orders of magnitude was observed in the temperature range from 0 to 150 K. The resistivity of $FeGa_3$decreases with increasing temperature which is also clearly semiconducting, and a large decrease in resistivity of more than seven orders of magnitude was observed in the small temperature range from 0 to 60 K.

The $FeGa_3$ is diamagnetic with susceptibility $\chi \sim -2*10^{-5}$ emu/mol, weakly temperature dependent between 150-350 K, with some upturn below 150K which is the development of a Curie tail that originates from tiny amount of impurities in the system. Magnetic susceptibility of $FeSb_2$ increases with an increase in temperature from a low temperature of 25*K* and passes through a region of diamagnetic to paramagnetic crossover and becomes paramagnetic at high temperatures. The crossover temperatures are around 125*K*. Minimum susceptibility values were found at approximately 50*K*. At T<25*K*, the development of a Curie tail can be associated with saturation from impurity states dominating the magnetic contribution in the sample once all intrinsic states are "frozen" out.

In general, n-type behavior with electrons as majority carriers is observed over the whole temperature range for Seebeck measurements on $FeGa_3$, the onset of increasing of Seebeck coefficient consistent with the resistivity decresing. Seebeck measurements on $FeSb_2$crystals showed behavior consistent with the activated behavior also evidenced by resistivity data, with the increase in thermopower below 25K, and peaking around 10K. Interestingly, a crossover of negative value to positive value was observed at about 6K which might indicate an n-type to p-type transition.

Reference

1 C. Petrovic, J. W. Kim, S. L. Bud'ko, A. I. Goldman,11 P. Canfield, W. Choe, and G. J. Miller, Phys. Rev. B. **67**, 155205 (2003).

2 Y. Hadano, S. Narazu, M. A. Avila, T. Onimaru, and T. Takabatake, J. Phys. Soc. Jpn. **78**, 013702 (2009).

3 A. Bentien, S. Johnsen, G.K.H. Madsen, B.B. Iversen, and F. Steglich, Europhysics Letters. **80**17008 (2007).

4. M. Wagner-Reetz, D. Kasinathan, W. Schnelle, R. Cardoso-Gil, H. Rosner, Y. Grin, Phys. Rev. B. **90**, 195206 (2014).

430203 Understanding the Configuration-Mechanical Stability Relationships for Si-CNT Heterostructured Anodes for Li-Ion Battery: A Computational Study

Tuesday, November 10, 2015: 8:30 AM
251D (Salt Palace Convention Center)

Sameer Damle, Chemical Engineering, University of Pittsburgh, Pittsburgh, PA, Siladitya Pal, Department of Mechanical and Industrial Engineering, Indian Institute of Technology Roorkee, Roorkee, India, Spandan Maiti, Bioengineering, University of Pittsburgh, Pittsburgh, PA and Prashant N. Kumta, Department of Bioengineering, University of Pittsburgh, Pittsburgh, PA

Understanding the configuration-mechanical stability relationships for Si-CNT heterostructured anodes for Li-ion battery: A computational study

Lithium ion batteries are one of the most sought after energy storage devices due to their relatively higher energy storage density, long cycle life, and an improved rate capability response as compared to other battery technologies such as Ni-Cd, Ni-MH, Lead acid, etc. Current state of the art Li ion batteries still implement graphitic anodes exhibiting a theoretical electrochemical capacity of 372 mAh/g. In the past decade, silicon has evolved as the most promising alternative anode material to graphite for Li ion systems due to its high theoretical gravimetric capacity of 4200 mAh/g. However, colossal volume expansion related mechanical degradation and eventual failure of Si during electrochemical cycling leading to loss in capacity limits the cycle life thus impeding the commercialization pathway.

Experimental as well as modeling studies have revealed that the use of nano-sized amorphous silicon (*a*-Si) morphologies significantly improves the anode capacity retention over multiple electrochemical cycles. However, *a*-Si suffers from poor electronic conductivity and charge transport, significant first cycle irreversible loss, inferior performance at higher current rates, and low columbic efficiency. On the other hand, multiwall carbon nanotubes are known to have very good mechanical strength along with excellent electrical and thermal properties. Thus, core-shell heterostructures comprised of carbon nanotube (CNT) core and nanostructured Si shell have emerged as promising candidates for Si based anode for Li-ion batteries. Different Si-CNT heterostructures ranging from uniform silicon thin films coated onto the CNT to Si nano-droplets tethered to the CNT at a specified spacing between the adjacent droplets can be synthesized. However, configuration of the Si component in the heterostructure is known to significantly alter the cycling performance of the electrode as shown by Epur et al. [2]. In this study, we focus on understanding the role of silicon configuration on the mechanical stability of the heterostructure during its electrochemical cycling. Such knowledge will shed light on contributing mechanisms for capacity fade in Si-CNT heterostructured anodes.

We hypothesize that nucleation and growth of voids in silicon during electrochemical cycling can induce its fracture and eventual failure. We utilize a custom developed multi-physics finite element (FE) modeling framework taking into account the Li diffusion induced elasto-plastic deformation of Si [2]. We systematically vary the geometry of Si in the Si-CNT heterostructure and simulate the resulting anode configuration for one complete electrochemical cycle (see Fig. 1). Comparison of the conditions for void growth and nucleation in the Si is done to understand the mechanical stability of the different Si-CNT heterostructured configurations during electrochemical cycling. Qualitative comparison of performed simulation studies with experimental results is made. Results from this study are expected to aid in the fabrication of improved Si/CNT heterostructure anodes. Results of the study will be presented and discussed.

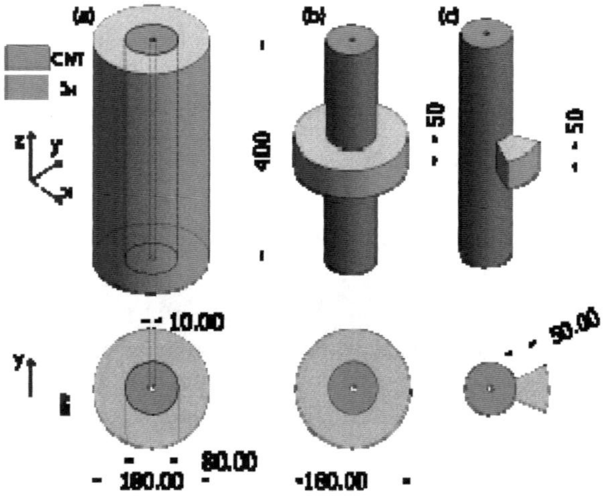

Fig. 1: Schematics of Si-CNT heterostructure geometries studied (a) Continuous Si film coating on CNT (b) Si nano-ring adhered to CNT and (c) $1/8^{th}$ of Si nano-ring adhered to CNT. Dimensions of the heterostructure components are specified in nm.

References:

1. Epur, R., M.K. Datta, and P.N. Kumta, *Nanoscale engineered electrochemically active silicon–CNT heterostructures-novel anodes for Li-ion application.* Electrochimica Acta, 2012. **85**: p. 680-684.

2. S. Pal, et al., *Modeling of lithium segregation induced delamination of a-Si thin film anode in Li-ion batteries.* Computational Materials Science, 2013. **79**: p. 877-887.

431072 A Fully Integrated, Efficient and Stable Solar-Driven Water-Splitting Prototype

Abstract:

Efficient, stable and scalable solar-driven water-splitting cells have the potential to store solar energy on the terawatt scale. The solar-hydrogen generator includes integration of the components for light absorption and charge separation, electrocatalysis for both hydrogen evolution reaction (HER) and oxygen evolution reaction (OER), and ionic transport between the two reaction chambers. In this talk, I will present recent advancement in the prototyping development of a fully integrated, wireless solar-hydrogen cell in Joint Center for Artificial Photosynthesis (JCAP). Guided by multi-physics modeling and simulation, a range of distinctive solar-hydrogen prototypes has been fabricated, evaluated and characterized from a stand-alone PV+electrolyzer system to a fully integrated, monolithic photoelectrochemical system. Specifically, a fully integrated, acid-stable and scalable louvered operational solar-driven water-splitting system has been modeled numerically and constructed experimentally. The louvered design maximized the area for light absorption and also provided necessary pathways for ion transport between the cathode and anode compartment. Leveraging the recent development of TiO_2 films for photoelectrochemical protection, un-assisted solar-driven waters splitting at solar-to-hydrogen conversion efficiency above 13% has been achieved in a fully integrated prototype for stable operation.

433815 Lithium Sulfide Cathodes Via Aerosol Spray Pyrolysis

Abstract:

The high promise of lithium-sulfur (Li-S) batteries attracts unprecedented interests in recent years. The majority of current investigations concentrate on sulfur-based cathode materials. The use of a sulfur cathode, however, presents several challenges due to the Li metal anode, which is prone to potential dendrite growth, shuttle reactions of Li with polysulfide species, and the reductive decomposition of electrolyte components on Li surface. One holistic strategy to address these problems is to produce batteries with Li_2S-based cathodes, with which high-capacity anodes including Si and Sn can be used instead of Li metal. The electrochemical reaction at the Li_2S cathode is as same as the sulfur cathode in principle. At the anode side, Li forms alloys with Si or Sn at higher potentials versus Li/Li^+ so that the Li dendrite formation can potentially be eliminated. The new batteries of Si-Li_2S and Sn-Li_2S indeed have lower energy comparing with Li-S batteries due to the lower cell voltage and the lower anode capacity. Nevertheless, they still have much higher specific capacity than the current Li-ion batteries.

The Li_2S cathodes also have advantages from the manufacture standpoint comparing with the conventional Li metal anodes (typically Li thin foils), which require sophisticated processes (purification, extrusion, passivation, etc.) due to the high reactivity of Li. On the other hand, Li_2S is nonflammable and oxygen stable (still need protection from moisture) so that the manufacturing process may be less demanding. In this study, we present a scalable process to synthesis Li_2S-carbon composite cathodes from various lithium salts and organic carbon precursors via aerosol-assisted spray pyrolysis (ASP). The ASP is a robust method to produce Li_2S-C composites with rational microstructures from different reactants. The reactions mechanisms and characterizations of the produced Li_2S-C cathode will be discussed in details.

431999 The Role of Polymer Composite Binder on Mechanics and Performance of Lithium Ion Battery Electrodes

Abstract:

Battery electrodes are complex multiphase composites which must provide efficient bicontinuous networks for transport of electrons (through the particle phase) and positive lithium ions (through the electrolyte filled pores of the electrode). A crucial but often neglected element of battery electrodes is the binder, typically a mixture of polyvinylidene fluoride (PVDF) and carbon black. The binder has two primary roles – to provide mechanical integrity and to improve electrical conduction of the electrodes. Migration of the binder has also been implicated as a potential mechanism of capacity fade in rechargeable lithium ion batteries.

We will present experimental characterization of the polymer binder for battery applications. Mechanical properties of the composite binder will be shown for both dry films and also binder swollen with carbonate electrolytes used in rechargeable batteries. Additionally we will examine the impact of mechanical stresses on the retention electrolyte and also the electrical conductivity.

Mesoscale simulations will be presented using experimentally determined three dimensional structures of battery cathodes. As batteries charge and discharge, the particles within the electrodes expand and contract as lithium is transferred into and out of the active material. The role of complex particle geometry and polymer binder properties on peak mechanical stresses due to particle swelling during cycling will be examined. Implications for battery performance and cycling stability will be discussed.

Sandia National Laboratories is a multi-program laboratory managed and operated by Sandia Corporation, a wholly owned subsidiary of Lockheed Martin Corporation, for the U.S. Department of Energy's National Nuclear Security Administration under contract DE-AC04-94AL85000.

433847 Electropolymerized Polyaniline/Manganese Iron Oxide Hybrids with Enhanced Electrochemical Energy Storage and Color Switching Response

Abstract:

Electropolymerized Polyaniline/Manganese Iron Oxide Hybrids with Enhanced Electrochemical Energy Storage and Color Switching Response

Yiran Wang,[1] Huige Wei,[1] Jiang Guo,[1] Suying Wei[2,*] and Zhanhu Guo[1,*]

[1]Integrated Composites Laboratory (ICL)

Department of Chemical & Biomolecular Engineering

University of Tennessee, Knoxville, TN 37909 USA

zguo10@utk.edu

[2]Department of Chemistry and Biochemistry

Lamar University, Beaumont, TX 77710 USA

suying.wei@lamar.edu

Polyaniline (PANI) nanocomposites embedded with manganese iron oxide ($MnFe_2O_4$) nanoparticles were prepared as thin films by electropolymerizing aniline monomers onto indium tin oxide (ITO) glass slides pre- spin-coated with $MnFe_2O_4$ nanoparticles. The shift of the characteristic peaks of PANI/$MnFe_2O_4$ in UV-visible absorption spectra and fourier transform infrared spectroscopy (FT-IR) indicate a formation of composite films and a chemical interaction between the PANI matrix and $MnFe_2O_4$ particles, which is also confirmed by the SEM images. A coloration efficiency of 206.2 cm^2 C^{-1} was obtained for the PANI/$MnFe_2O_4$ nanocomposite film, higher than that of the pristine PANI film, 104.2 cm^2 C^{-1}, suggesting a synergistic effect between the $MnFe_2O_4$ particles and the PANI matrix. An enhanced areal capacitance as 4.46 mF cm^{-2} was also achieved in the PANI/$MnFe_2O_4$ nanocomposite film compared with that of 3.95 mF cm^{-2} in the pristine PANI film from the CV at a scan rate of 5 mV s^{-1}. The enhanced capacitance of the composite films are attributed to the pseudocapacitive property of $MnFe_2O_4$ and the rougher morphology caused by the embedment of $MnFe_2O_4$ particles into the PANI matrix. Finally, the sulfuric acid (H_2SO_4) concentration and temperature effects on the supercapative behavior of the pristine PANI and its $MnFe_2O_4$ nanocomposites films were studied, suggesting positive effects of decreasing H_2SO_4 concentration and increasing temperature during a low temperature range, relative higher temperatures can destroy the PANI structure and cause the degradation of PANI.

1. H. Wei, C. He, J. Liu, et al, Polymer, doi:10.1016/j.polymer.2015.04.064.
2. H. Wei, Y. Wang, J. Guo, X. Yan, et al, ChemElectroChem, 12(1) (2015) 119-126.
3. X. Zhang, J. Zhu, N. Haldolaarachchige, et al, Polymer, 53(10) (2012) 2109–2120.

433261 Performance Metrics and Design Principles for Cost-Effective Separators in Non-Aqueous Redox Flow Batteries

Abstract:

In a range of electrical energy storage technologies, redox flow batteries (RFBs), which reversibly convert chemical energy to electrical energy, have shown a favorable balance of cost, safety, and performance for stationary applications. Shifting from aqueous to non-aqueous electrolytes may promise a higher cell voltage due to the extended electrochemically stable window (~ 4 V) and an enriched selection of redox materials due to the broader variety of organic solvents. As a key component in RFBs, an ideal membrane should have negligible electronic conductivity, high ionic conductivity, good chemical stability and mechanical strength, and low active species crossover. However, to date, no non-aqueous membrane has been identified that simultaneously fulfills these requirements, especially high ionic conductivity and low crossover rate.

To meet aggressive established cost targets ($120/kWh), our recent work suggests the maximum area-specific resistance for non-aqueous RFBs is 2.3 $\Omega\cdot cm^2$, which corresponds to the ionic conductivity of 1.1 mS/cm and 7.7 mS/cm for a membrane with a thickness of 25.4 μm (such as Nafion® 211) and 177.8 μm (such as Nafion® 117), respectively. However, there is little data on the effectiveness of successful aqueous membranes under non-aqueous conditions. To this end, we explore the performance of ion-selective lithiated Nafion 117 (Li N117) and non-ion selective nanoporous separators with a range of well-defined pore sizes in non-aqueous electrolytes with a focus on the relationships between ionic conductivity, species selectivity, and membrane microstructure (pore size). Key membrane properties and performance metrics are quantified in electrolytes consisting of different salt and solvent combinations, and compared between ion-selective membranes and non-selective separators. Results indicate a critical relationship between pore size, ionic conductivity and species selectivity and shed light on the future research directions in non-aqueous separations.

The Performance of Structured High-Capacity Si Anodes for Lithium-Ion Batteries

15AIChE
Annual Meeting, Salt Lake City, UT

Department of Chemical Engineering*
Department of Physics and Astronomy†
Brigham Young University, Provo, Utah
J. Fan*, L. Barrett†, S. Palmer*, R. C. Davis†, R. R. Vanfleet†, J. N. Harb*

MOTIVATION FOR THE RESEARCH

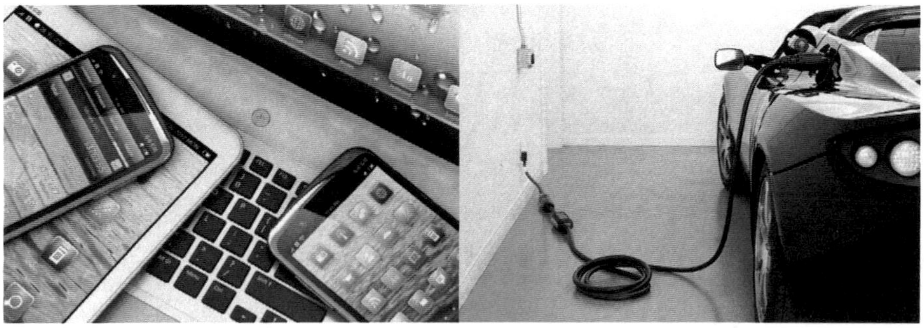

2

WHY SILICON?

1. Fully-lithiated Li_5Si_4 alloy (3,579 mA/g) has 10 times higher capacity than LiC_6.
2. Si has a low operating potential (0.4–0.6 V vs. Li/Li^+).
3. Abundance in earth soil

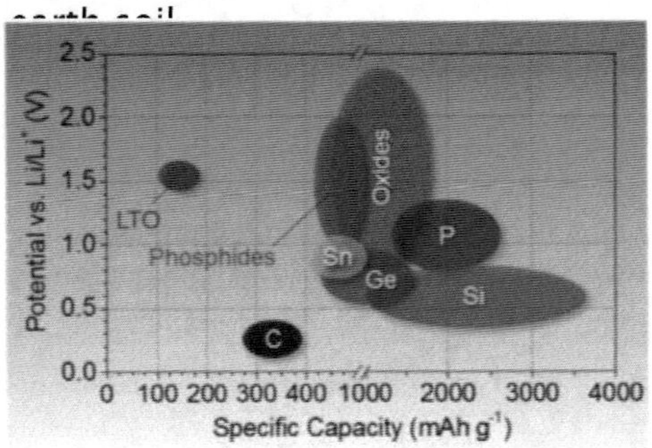

N. Nitta, et al., Mater. Today (2014), 18(5), pg. 252-264

3

TWO IMPORTANT ISSUES OF SI ANODE:

1) Large volume changes during cycling

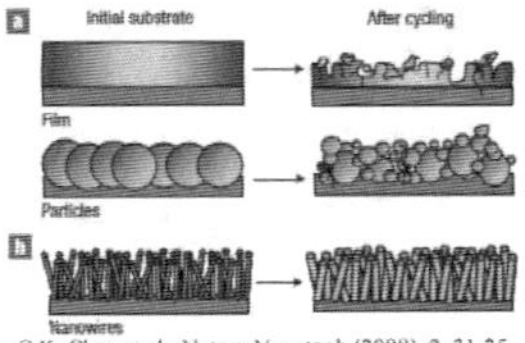

C.K. Chan et al., Nature Nanotech (2008), 3, 31-35

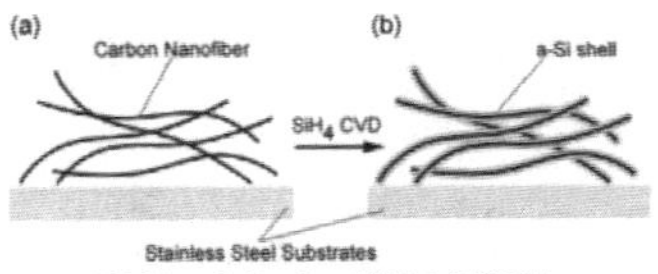

L.F. Cui, et al., Nano lett., (2009), 9, 3370-3374

(2) Formation of solid-electrolyte interphase (SEI)

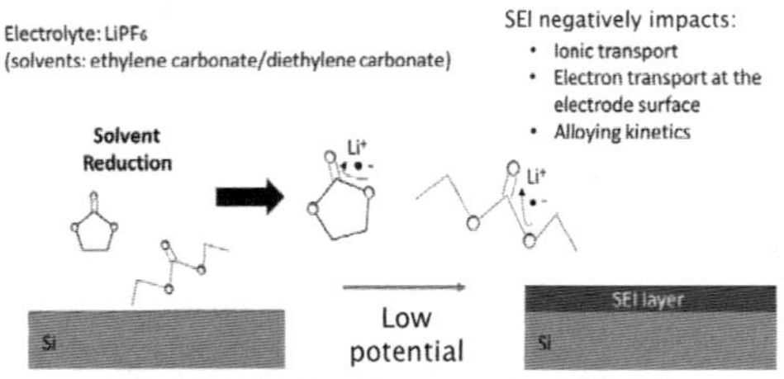

Key SEI-related Issues
- Large surface area for SEI formation in nanostructured electrodes
- Stability of SEI on Si anode

4

MICROFABRICATION OF SI-VACNT ELECTRODES

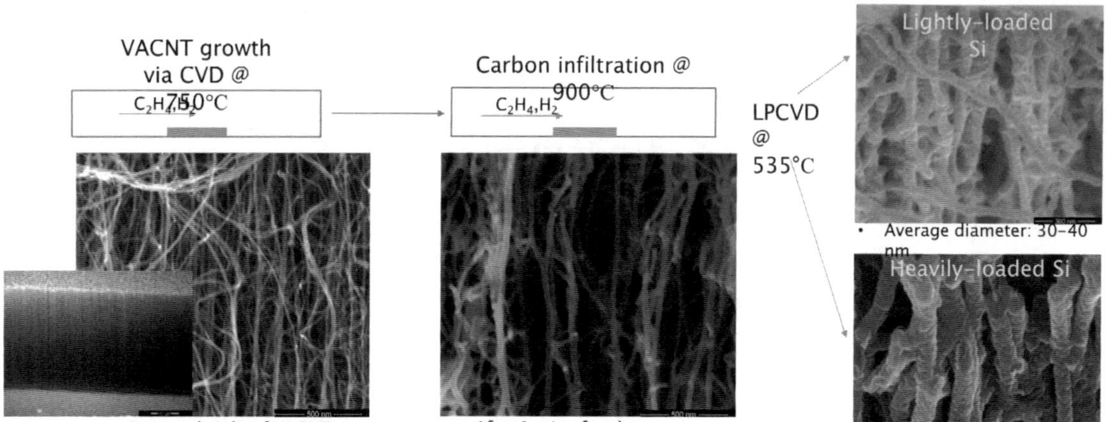

- Average height of VACNTs : ~70 μm (1 min) or ~135 μm (2 min)
- Average diameter: 11 nm

- After 1 min of carbon infiltration, the average diameter is about 25 min

- Average diameter: 40-50

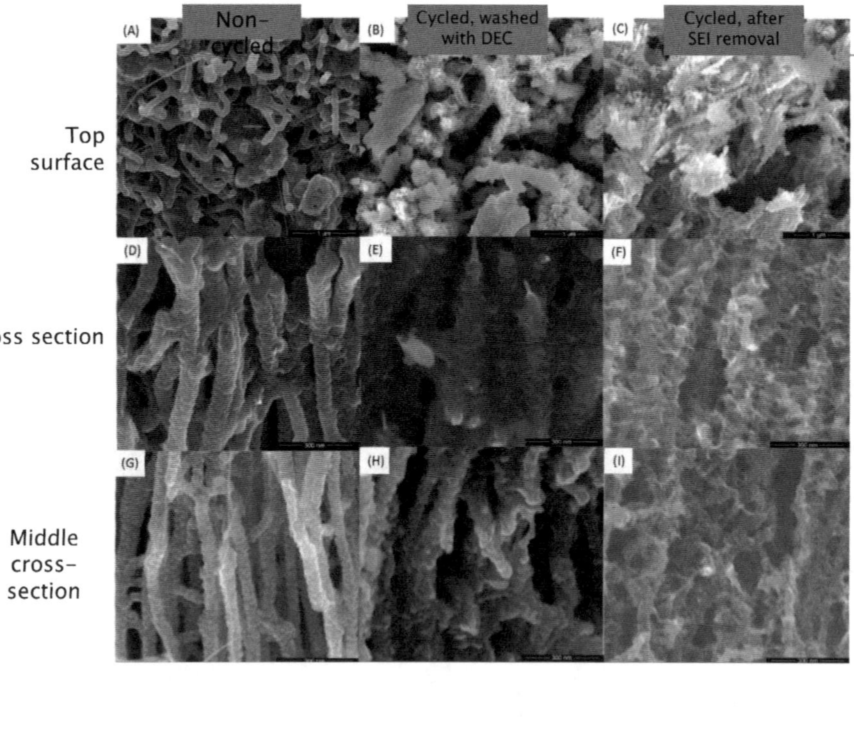

6

ENCAPSULATION PROCESS

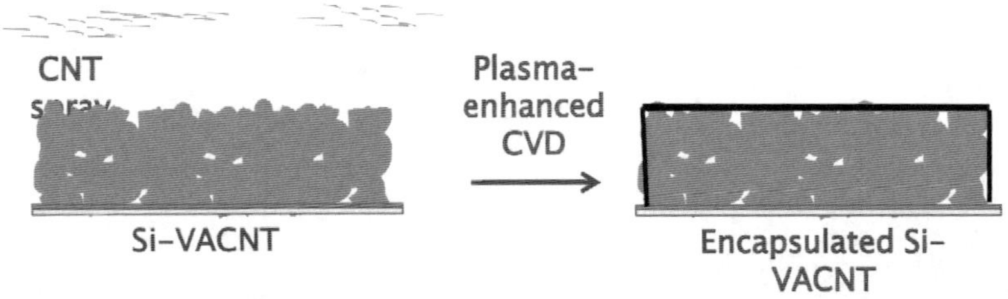

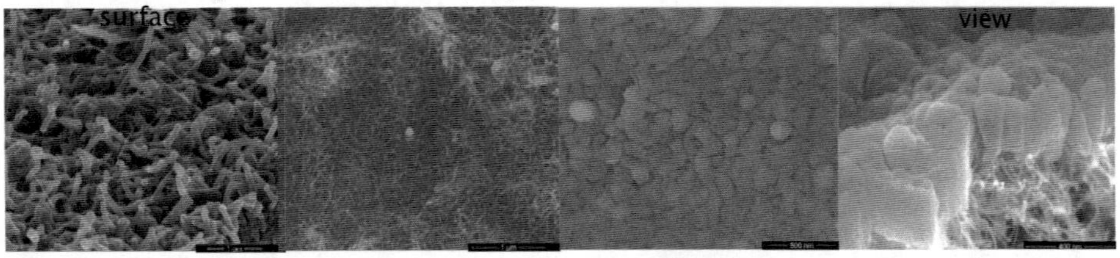

7

POST-CYCLING CHARACTERIZATION OF THE ENCAPSULATED ELECTRODES

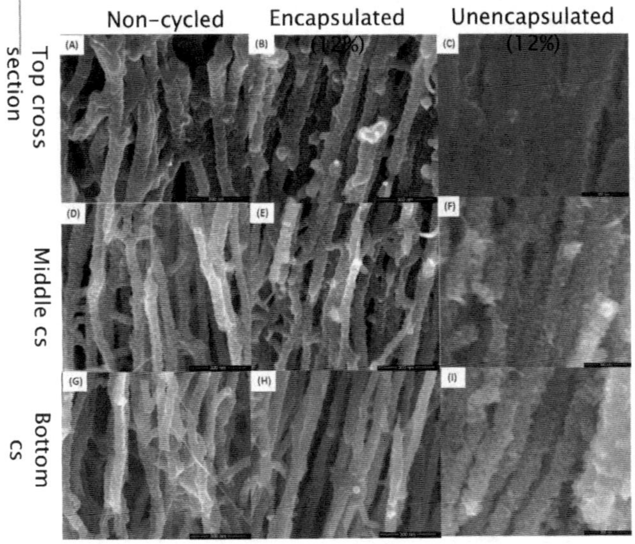

12% Si EDX comparison

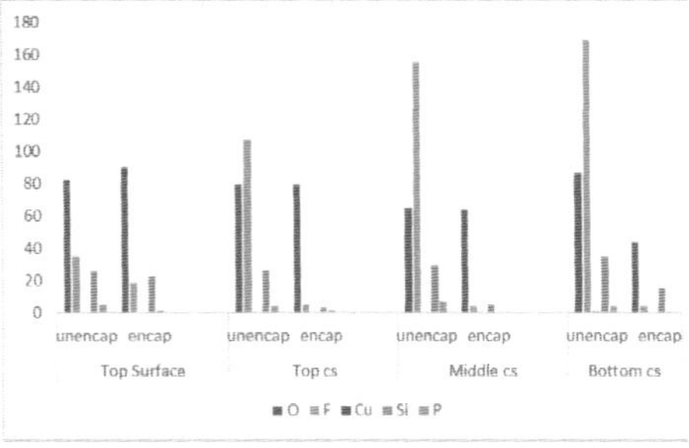

8

CYCLING PERFORMANCE

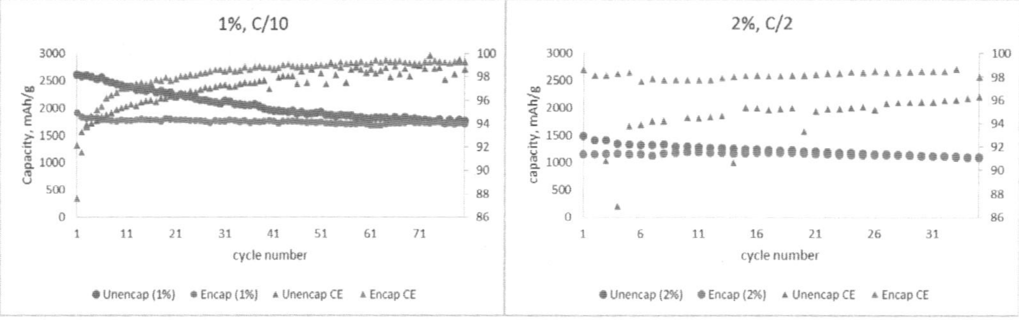

LI TRANSPORT IN ENCAPSULATED ELECTRODES

Dimensions:

- Si-VACNT diameters: heavily-loaded is 40-50 nm and lightly-loaded is 30-40nm.
- Si-VACNT height: 70 (1min)-135 (2 min) μm.

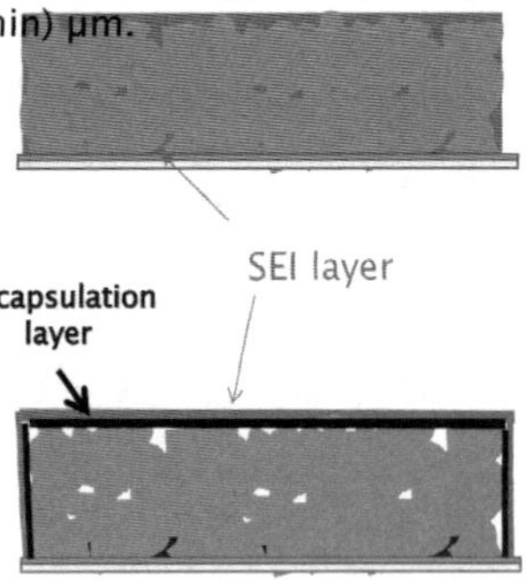

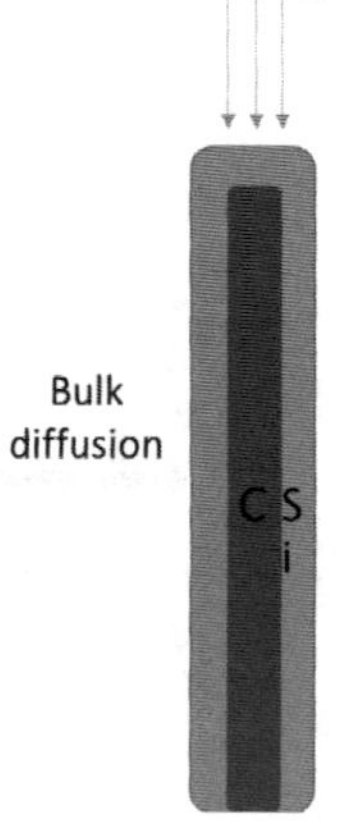

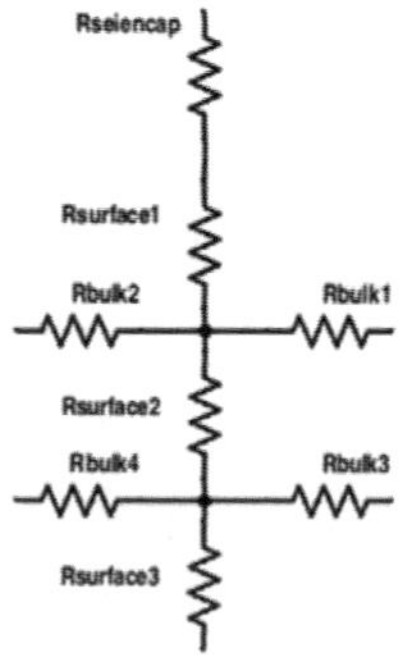

10

Q: DOES ELECTRODE HEIGHT LIMIT LI TRANSPORT DOWN THE TUBES?

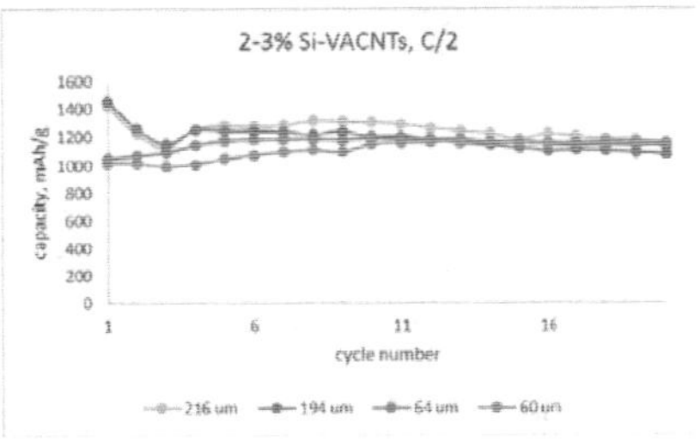

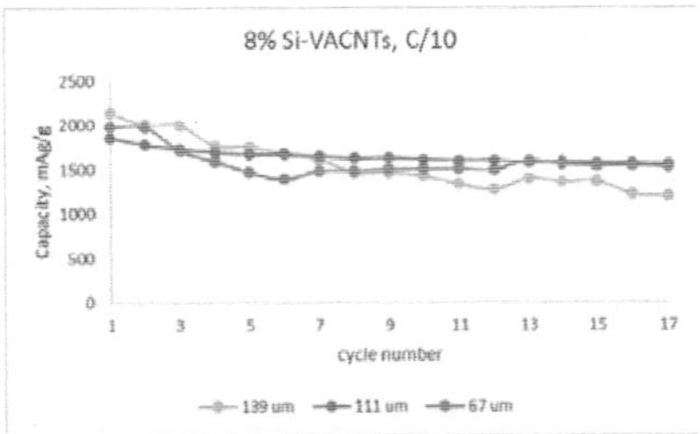

Experimental Design:

- Cells with similar Si% but various heights were cycled at C/2 or C/10
- If electrode height limits Li transport, the short electrodes would perform better
- No significant difference with height: *electrode height does NOT limit Li transport*

SAME SUPERFICIAL CURRENT DENSITY

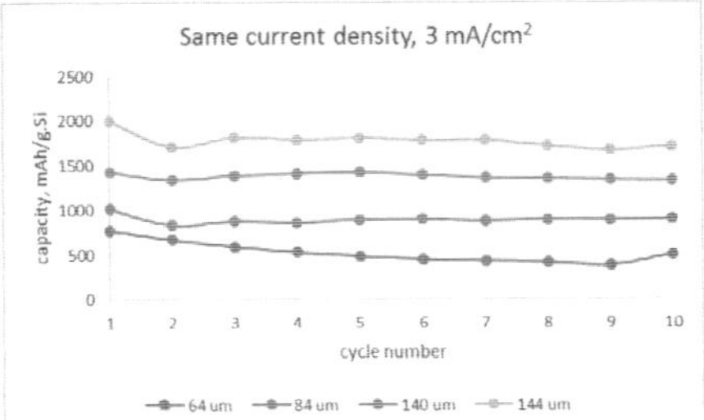

Experimental Design:

- Cell with similar Si% but different heights were cycled at the same superficial current densities as shown
- Tall electrodes performed better because they have relatively low interfacial current density.

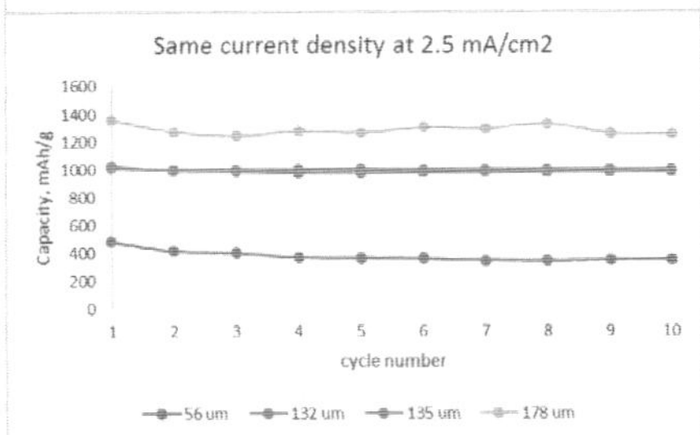

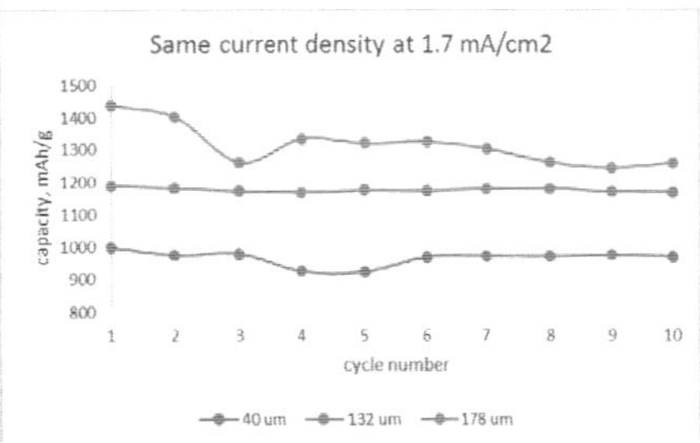

Q: WHAT'S THE ROLE OF ELECTRODE HEIGHT IN LI TRANSPORT?

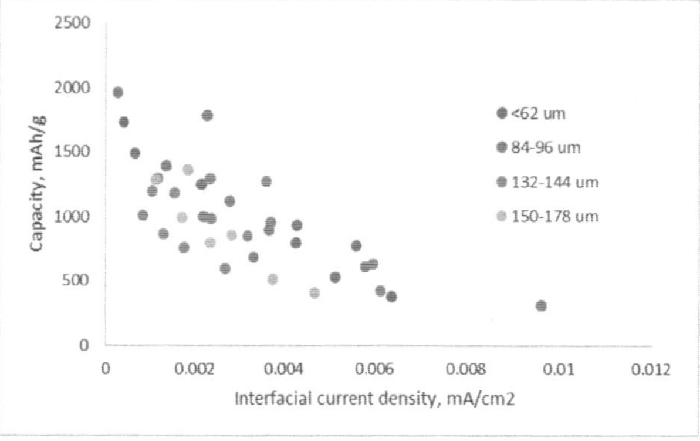

BULK DIFFUSION DOMINATES?

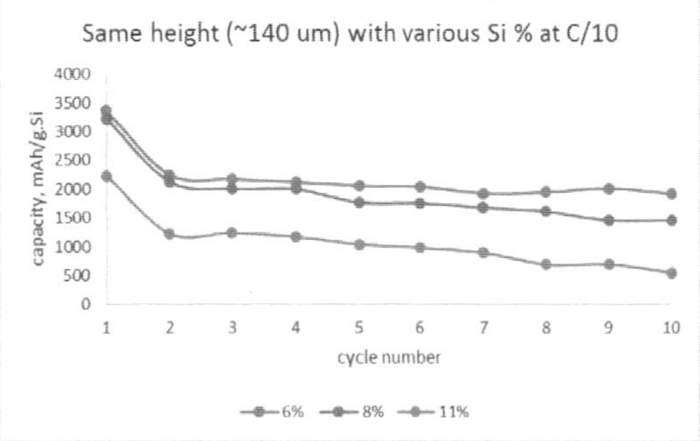

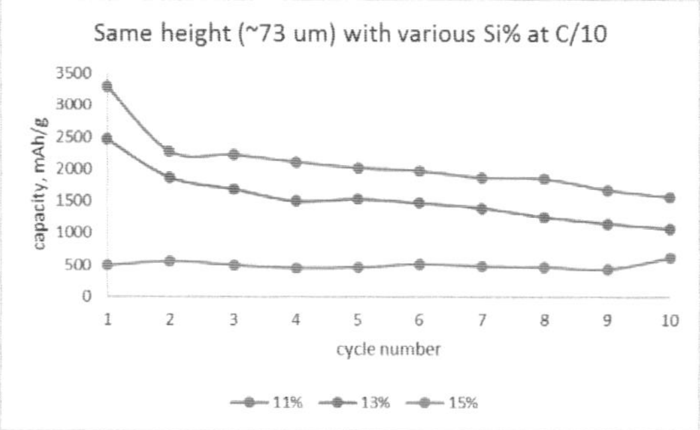

CONCLUSIONS

- The Si-VACNT template was aimed to study two key issues of Si-based anodes for lithium ion batteries: 1) accommodation of silicon volume changes during cycling and 2) the role of SEI formation in the Si-VACNT forests.
- The Si-VACNT electrodes without the encapsulation were filled up with SEI formation products after long cycling. Encapsulated Si-VACNT electrodes tend to experience less morphological changes during the electrochemical testing.
- Li transport, presumably due to surface diffusion, did not appear to limit cell behavior, even when height was orders of magnitude greater than the tube diameters. Therefore, encapsulation may be a viable strategy for minimizing the amount of SEI and improving electrode performance.

15

THANK YOU

the National Science Foundation (CBET-116028

431083 Graphene Oxide Based Foams for Lithium Ion Batteries

Abstract:

Advances in science and engineering related to emerging technologies for lithium-ion batteries (LIBs) have been so spectacular in recent years that they have become the most popular source for portable computing and telecommunication equipment. In the last five years there has been a lot of work on the electrochemical properties of electrode Materials for Lithium Ion Batteries (LIB) using Graphene (G) and Graphene Oxide (GO) with silicon , however, these works present several drawbacks associated with the breakage of the silicon layer. We present a new technology that avoids breakage of the silicon layer at the SEI interface and their characterization. We incorporate silicon as the high capacity electrochemically active material which has the potential to greatly increase the capacity of these materials. We thermally reduce these composites to convert GO to a graphitic structure providing high electrical conductivity as well as high lithium ion conductivity. The uniform cellular structure of these foams provides a method to obtain a uniform dispersion of the active material inside the composites. We solidify the composite to effectively trap the electrochemically active nanoparticles within the foam preventing segregation of the active components from the conductive matrix. This work also focuses on using the matrix to create a barrier between the active component and the electrolyte. Such a barrier, even only a few nanometers in thickness, has the potential to limit the inefficiencies created by the breaking and reformation of the solid electrolyte interface upon cycling. This allows high capacity active materials to be used with improved cycling stability, efficiency, and power. Our results show anodes cycled at very high charge and discharge rates of 3A/g are possible. At these rates capacities over 800 mA-hr/g were demonstrated at 40 cycles.

(286i) Solubility Product Observations in Formation of Active LiFePO4 Cathodic Material for Secondary Energy Storage Application

Kwee, D. W., University of California Riverside
Martinez-Morales, A. A., University of California Riverside
Originally presented on: 11/10/2015 10:30:00 - 10:45:00

An Ionic Liquid medium has been a promising alternative to solvent-based synthesis for a variety of materials due to its attractive wide temperature window and scalability potential. Ionic Liquid medium for use in synthesis of crystalline material for active cathodic material had been recently explored, though precursor solubility remains a long-time issue. In this approach, water is initially used as solvent for formation of intermediates, enabling effective precursor solubility. Ionic liquid medium is then used to drive the reaction, from intermediate (created separately from water-mediated reaction) to product. Reported here are solubility product Ksp observations for the solubilization / reprecipitation of intermediates along with its direct effect on active $LiFePO_4$ product crystallinity, morphology, and electrochemical performance.

433105 Understanding Lithiation Mechanisms in Silicon-Based Nanomaterials from First Principles

Abstract:

Silicon-based alloys, composites and nanostructures have received great attention as a possible replacement of conventional carbon-based anodes due to their higher lithium storage capacity. However, still fundamental aspects of lithiation processes and properties in Si-based nanomaterials remain largely ambiguous, despite their importance in overcoming many technical hurdles faced in practical use. Using first principles-based atomistic modeling, we have explored lithiation mechanisms in various Si-based nanosystems including nanowires, composites, oxides, and alloys. In this talk, we will present our recent progress in understanding the impacts of surfaces, interfaces and alloying elements on the properties and performance of Si-based anode materials for Li-ion batteries. Our first principles study shows that the lithiation behavior of Si nanostructures and composites is considerably different from the case of bulk Si because of surface and interface effects. For instance, the mobility of Li along the surface or interface tends to be significantly enhanced by several factors. We also find that by alloying Si with certain metal (M) elements, the theoretical capacity is compromised slightly, but in return the Si-M network may help stabilize the lithiated host matrix and therefore contributes to the improved cycling performance. We will also briefly touch on the importance of alloy composition and local atomic environment in determining the lithiation properties and performance. The improved understanding can contribute to the rational design of Si-based anode materials to maximize capacity retention and rate performance.

422604 Novel Centrifugation-Assisted Preparation (CAP) of Additive-Free Carbon-Decorated Fe3O4 Electrodes with Superior Electrochemical Performances for Lithium-Ion Batteries

Abstract:

2015 AIChE Conference

Novel Centrifugation-Assisted Preparation (CAP) of Additive-Free Carbon-Decorated Fe_3O_4 Electrodes with Superior Electrochemical Performances for Lithium-ion Batteries

Jian Zhu (speaker); K. Y. Simon Ng; Da Deng*

Department of Chemical Engineering and Materials Science, Wayne State University,

5050 Anthony Wayne Dr, Detroit, MI, United States, 48202

Email: da.deng@wayne.edu

Abstract

Lithium ion batteries (LIBs) are the dominant power sources for portable electronic devices, and are penetrating into the market of electric vehicles and sustainable energy grid. There is an increasing demand for higher energy and high powder LIBs. The electrochemical performances of LIBs are mainly determined by the electrodes selected. Taking anode for example, the theoretical capacity of commercial anode of graphite is only 372 mA h/g, and its performance is about to saturate after few decades of development. To further increase the energy density, carbon must be replaced by environmentally friendly and inexpensive materials with high capacity and excellent reversibility. On the other hand, the current conventional procedure in the preparation of electrodes is time consuming involving multiple steps, including materials preparation, slurry formation, coating on to current collectors, and drying in vacuum. Various additives, including binder polyvinylidene fluoride (PVDF), conductive enhancer carbon black (CB), and organic solvent N-Methyl-2-pyrrolidone (NMP), are required in electrode preparation. The complicated procedure and various additives lead to the high cost of electrodes. It would be beneficial but is still challenging to develop facile methods for electrode preparation with simplified procedure, without using any additives, but with superior electrochemical performances.

Herein, we demonstrated a novel and facile centrifugation-assisted preparation (CAP) of additive-free negative electrode of carbon-decorated Fe_3O_4 materials for high energy LIBs (Figure 1a). Notably, the materials were directly synthesized on the current collectors. No additives, such as PVDF binder, carbon black conductivity enhancer and NMP solvent which are typically used in conventional electrodes, were required. Interestingly, the as-prepared of carbon-decorated Fe_3O_4 negative electrodes exhibited superior electrochemical performance with stable specific capacity of 800 mAh/g (more than two times of theoretical capacity of graphite at 372 mAh/g) for at least 240 cycles at 100mA/g; (Figure 1b) no significant capacity decay was observed when the testing currents were increased, up to 1500 mA/g, demonstrating excellent rate performances for high power LIBs (Figure 1c). We believe that our facile CAP method could be used for the preparation of electrodes, including cathodes and anodes with or without additives, based on other active materials.

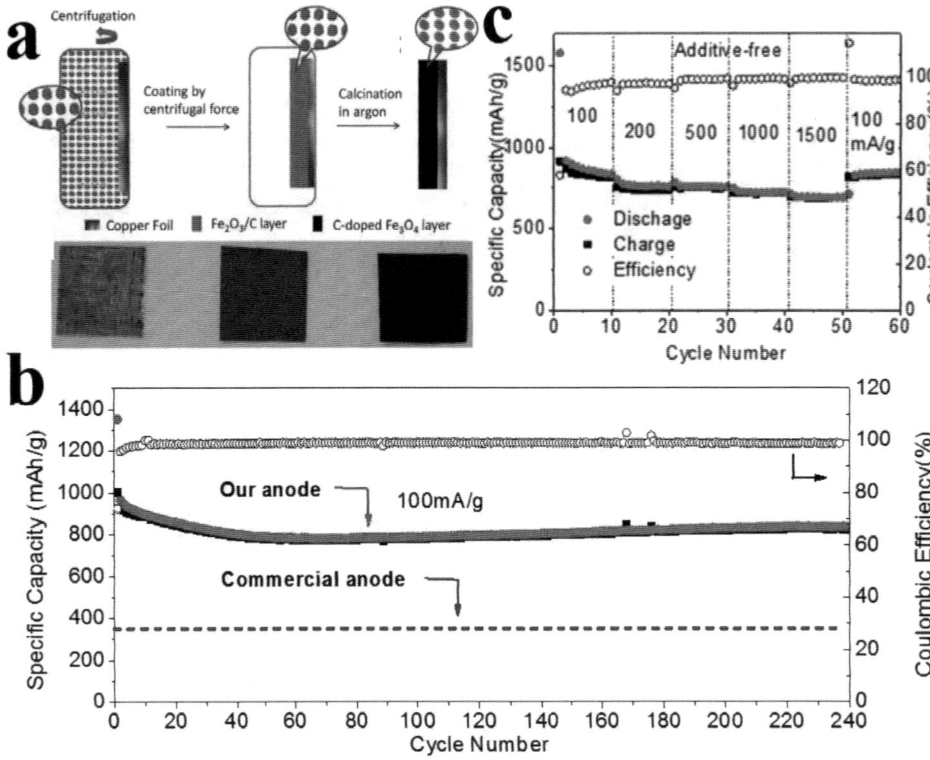

Figure 1. (a) Schematic of three steps of CAP of additive-free electrodes with carbon-decorated Fe_3O_4 on Cu current collector and corresponding optical images on the bottom; (b) excellent cycling performance tested at current of 100 mA/g for 240 cycles; and (c) rate performances and Coulombic efficiency at different currents from 100 to 1500 mA/g of the CAP electrodes of carbon-decorated Fe_3O_4 without any additives.

424677 Stacked SnS2/Graphene Nanocomposites with High Li+ Storage Capacity

Abstract:

There is strong recent interest in using SnS_2/graphene composites for reversible Li^+ storage. While SnS_2 and graphene are 2D nanomaterials; many of the composites reported thus far have a random 3D structure. We prepared a stacked SnS_2/graphene nanocomposite which could retain most of the properties associated with the original 2D nanomaterials and promote the reversible storage of Li^+ via the conversion reaction ($SnS_2+4Li^++4e^-\rightarrow Sn+2Li_2S$). Contrary to the prevailing view that the conversion reaction is irreversible, we will show that reversibility as high as 87.5% is attainable through careful materials design. The increased reversibility also imparted to this composite specific capacity, capacity retention and rate performance which surpassed previously reported SnS_2/graphene composites measured under the same conditions.

431500 Gas-Assisted Electrospray for Ultrauniform, Ultrafast, Instantly-Dry, and Binder-Free Electrode Preparation

Abstract:

Lithium-ion batteries has been a hot research focus in recent years due to their high energy density, flexible and lightweight design, long lifespan, and wide application in portable electronic devices and electric vehicles. Among all the battery-related researches, most focus on electrode materials, a small portion focus on binder, separator, electrolyte, and almost none focus on the preparation of electrode. For the tons of researches, materials preparation could vary a lot, but almost all used either drop cast or slurry coating for final electrode preparation. However, both drop cast and slurry coating methods involve toxic organic solvents and consist of multi steps including mixing, sonication, blading, calendaring, and long-time vacuum drying. Even after all these tedious steps, the as-prepared electrodes still suffer from exfoliation of the active material from current collector, leading to poor cycling stability. Though there are also some free-standing electrodes skipping the traditional electrode preparation procedure, they are not versatile and are often limited to certain kinds of materials.

To simplify the preparation process, and more importantly get better adhesion and high uniformity, we have developed a facile electrospray process from the already industrialized electrospining process. It can be applied to spray slurry (binder, additive carbon, active materials) from both organic and water system. The as-sprayed electrodes not only show no exfoliation upon bending, but also have lower resistance demonstrated by smaller impedance circle in comparison with conventionally made electrodes. The flow rate is 10 or 20 times fast of electrospinning, thus desired mass loading of electrodes can be achieved with only a few minutes of spraying. Excitingly, the electrodes made can be directly used for battery assembly, no extra drying step required. We further advanced the technique to a general binder-free electrode preparation approach where only graphene oxide is added. The graphene oxide sheets not only act as binder and conductive agents, but also act as "pocket" to hold active materials and help resist mechanic stress during charge and discharge, giving outstanding battery stability. In addition to battery electrodes, this approach can be also applied for many other electrode preparations such as supercapacitors, fuel cell, and solar cell etc.

428976 SnO2/Carbon Nanotube–Infiltrated Ni Nanofoams As 3D Anodes for Enhanced Performance of Lithium-Ion Batteries

Abstract:

Metallic nanofoams have been widely investigated as 3D current collectors in lithium-ion batteries (LIBs) due to their high stability and mechanical robustness, as well as the highly conductive 3D framework that can be coated with high-capacity anode materials. However, the extremely high porosities (> 99%) and low specific surface areas (< 10 m^2/g) of metallic nanofoams, limits the amount of active electrode material that can be deposited, thus reducing the maximum achievable areal energy density. Although the weight penalty associated with metallic nanofoams is a disadvantage, their highly conductive interconnected fibers and high void fraction could be exploited for improved LIB performance by infiltrating the nanofoam with a highly conductive, low-density material such as carbon nanotubes (CNTs). Here, we report for the first time, an efficient and scalable gas-phase liquid injection CVD approach using a ferocene/xylene mixture for the infiltration of arrays of high-quality CNTs in pretreated Ni nanofoam. Importantly, the BET specific surface area of the pretreated Ni nanofoam after CNT infiltration increases by a factor of over 20. Further, upon deposition of SnO_2 on the CNT-infiltrated Ni nanofoams by atomic layer deposition, the new nanocomposites showed improved energy capacity and excellent rate capability, as well as high capacity retention during electrochemical charge-discharge measurements.

435122 Synthesis of Layer-By-Layer Thick Mesoporous Titania Films with Vertically Oriented 2D-HCP Nanopores and Their Use in Lithium Ion Batteries As Negative Electrodes

Abstract:

Mesoporous titania films with 2D Hexagonal Close Packed (HCP) cylindrical nanopores have been synthesized by the evaporation-induced self-assembly (EISA) technique with Pluronic surfactant F127 (with average structure $(EO)_{106}(PO)_{70}(EO)_{106}$ where EO is an ethylene oxide unit and PO is a propylene oxide unit) as the structure-directing agent. To provide vertical alignment of the pores, surface modification of substrates with crosslinked surfactant F127 has been used to provide a chemically neutral surface. According to Koganti et. al (***Nano Lett.*** 2006, 6, 2567), thick films do not have perfect vertical alignment of the pores and better vertical HCP orientation is best achieved by sandwiching the films between two modified surfaces. Here, we report the synthesis of relatively thick films with vertically oriented 2D HCP cylindrical nanopores by layer-by-layer synthesis. Such films do not require sandwiching for vertical orientation of the pores. The hypothesis being tested is whether surfactant micelles from an existing layer will interact with micelles from a freshly deposited layer to provide epitaxial self-assembly of vertically oriented cylindrical micelles in registry from layer to layer. Using this technique, titania films with vertically oriented cylindrical nanopores as much as 1 μm thick were obtained. The thickness and vertical orientation were characterized using XRD, SEM and profilometry. These films were then used as negative electrodes in Li-ion batteries because of advantages of titania including good cycling stability, small volume expansion (~3%) during intercalation/extraction and high discharge voltage plateau. The high surface area and small wall thickness of these titania films provide better lithium ion insertion and reduced Li ion diffusion length compared to nonporous titania films or titania nanotube arrays. However, maintaining accessibilty of the full pore space in the thick films is a challenge to be discussed. Furthermore, the mass transport across these films is studied using electrochemical impedence spectroscopy (EIS). Using equivalent circuit modeling, the fraction of the accessible area of the nanoporous film coated electrode was calculated and analyzed as a function of the number of layers deposited.

(306g) Engineered Ionic Diffusion Layers to Increase Rate Capability of NCA Cathode in Lithium-Ion Cells

Dahlberg, K., University of Michigan
Mohanty, D., Oak Ridge National Laboratory
Mahajan, V., XALT Energy, LLC
Lee, M., XALT Energy, LLC
Stevenson, L.,
Stanley, J., XALT Energy, LLC
King, D., PneumatiCoat Technologies
Wood, D. III, Oak Ridge National Laboratory
Albano, F., XALT Energy, LLC
Dhar, S., XALT Energy, LLC
Originally presented on: 11/10/2015 10:30:00 - 10:50:00

High energy and long cycle life are the key technical attributes required for widespread implementation of high-performance electric vehicles (EVs). High rate capability is the key to battery performance for many other applications including stationary grid connected applications and Hybrid Electric Vehicles (HEV's). Despite significant strides made to-date, the current high energy density solutions suffer from degradation and limited cycle life limiting the widespread implementation of NCA cathode materials in Li-Ion batteries. There is growing evidence that several degradation mechanisms influencing these limitations take place at the cathode surface, motivating the use of larger particle sizes to minimize surface area and mitigate degradation. Larger particle sizes, however, decrease the overall rate of diffusion of lithium ions, causing a loss of activation of Li-rich phases and an increased loss of capacity at higher rates. In this paper, we show that Al_2O_3, TiO_2, and other similar ALD coatings applied to NCA powder, continuously covering active material particle surfaces, create an engineered SEI layer which facilitates an improved Li-ion diffusion pathway, increasing the rate capability of NCA cathodes even with larger particle sizes. Electrochemical analyses showed increased rate capability in 95x64 mm pouch cells (~2.5Ah) with coated NCA compared to uncoated, and XRD and TEM were used to characterize the underlying favorable interactions of coatings with the cathode surface. XALT Energy's integrated cell design and manufacturing makes 95x64 mm cell performance representative of large format (216x216 mm) production EV cells, demonstrating the validity and scalability of this approach.

Relationship of Graphene Microstructrue and Ultracapacitive Property

Zhiyuan Xiong, Xiaogong Wang
Tinghua University, Beijing, China

Introduction

Since the pioneering work by Novoselov et al., graphene has sparked enormous research interest in scientific and industrial fields. The excellent electrical, electronic and mechanical properties as well as extremely large surface areas of graphene make it a competitive candidate for many applications. One of the promising applications is to be used in the electrochemical devices for the low-cost and environmentally-friendly energy storage.[4] Electrochemical capacitors (ECs), also known as supercapacitors, have been widely used in energy storage fields.[5] ECs can combine the high power performance of traditional capacitors and large energy density of batteries, which have played an important role complementary to batteries in the energy storage field. Depending on the charge storage mechanism, ECs can be classified into two categories, namely electrochemical double-layer capacitors (EDLCs) and pseudocapacitors.[6] Pseudocapacitors are fabricated by using Faradic pseudocapacitor materials, such as conducting polymers and various transition metal oxides. The energy storage is realized by fast and reversible electrochemical redox reactions between electroactive electrode materials and ions in suitable potential windows. On the other hand, EDLCs mainly use the active carbon with high surface areas as electrode materials and employ charge separation at the electrode/electrolyte interface to store charge.[7] Graphene as a two-dimensional material with an extremely high surface area is considered to be promising as the electrode material for EDLCs.[8-10] As the chemically converted reduced graphene oxide can be produced with the ton-scale yield at a low cost, this easy availability makes it possible for graphene to be used as practical electrode materials.[11]

At the current stage, the benefits of the graphene for EDLC applications have not been fully reaped. Theoretically, EDLCs derived from graphene can reach a high capacitance of 550 F/g if the surface areas of graphene can be made full of use.[12] However, it is almost inevitable for graphene to undergo the phase separation and irreversible agglomeration in the devices, which results in the loss of its surface areas as well as the reduction of the specific capacitance.[13,14] Preparation of graphene electrode with reasonable microstructures is significant and beneficial for us to tackle the challenge. There different microstructures, namely layer-by-layer (LbL) self-assembled multilayer films, coagulation self-assembled graphene macropore network and hydrogel post-reduced hierarchical porous graphene films were designed and developed for supercapacitor application.[15-17]

Layer-by-Layer self-assembled SGO/PSDAS film

Photocrosslinkable multilayer films composed of sulfonated reduced graphene oxide (SRGO) and polystyrene-based diazonium salt (PSDAS) were fabricated by electrostatic layer-by-layer (LbL) self-assembly.[18] Polystyrene with narrow molecular weight distribution was synthesized by atom transfer radical polymerization (ATRP) and cationic PSDAS was prepared through nitrification, reduction, diazotization reactions. Negatively charged SRGO nanosheets were prepared through pre-reduced by $NaBH_4$, modified by aryl diazonium salt of sulfonic acid, and then further reduced by hydrazine. The multilayer films were obtained by alternately dipping substrates in the PSDAS solution and SRGO dispersion in acid conditions (Scheme 1). The crosslinking between the components occurred during the multilayer formation process and was further achieved by the UV light irradiation after the film preparation. The assembling process and surface morphology of LbL multilayer films were monitored by UV-vis spectroscopy and Atomic Force Microscopy (AFM). The crosslinking between SRGO and PSDAS was verified by attenuated total reflectance FTIR (ATR-FTIR) spectroscopy, X-ray photoelectron spectroscopy (XPS), and contact angle measurement. The graphene nanosheets were found to be homogeneously distributed in the crosslinked network of the films. The large accessible surface area of graphene nanosheets and the crosslinking structure afforded the LbL films with high specific capacitance and excellent cyclic stability when used as supercapacitor electrodes (Figure 1). At a sweeping rate of 10 mV/s, the film with nine-bilayers exhibited a specific capacitance of 75.2 F/g with ideal rectangular cyclic voltammogram. Large capacitance retention of 97% was observed after 10000 charge-discharge cycles under the scanning rate of 1000 mV/s. This new approach for preparing graphene-containing multilayer films can be used for developing supercapacitor electrodes and other functional devices.

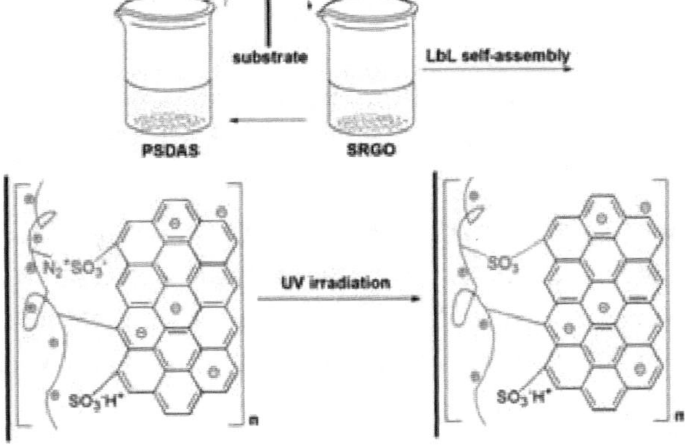

Scheme 1. Multilayer film prepared from layer-by-layer deposition and post-preparation crosslinking.

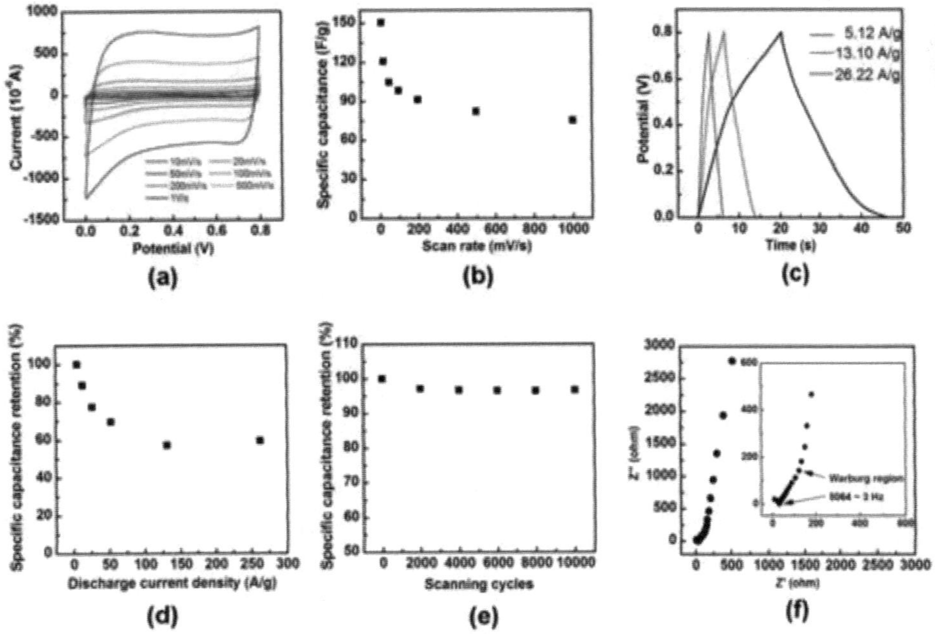

Figure 1. Electrochemical properties of the multilayer film as the electrode in 1.0 M Na2SO4 solution. (a) CV curves, (b) specific capacitance of multilayer films with 9 bilayers under different scanning rates, (c)

Galvanostatic charge/discharge curves at different current densities, (d) specific capacitance retention of the multilayer film with 9 bilayers versus the discharge current density, (e) Cyclic test of multilayer films with 9 bilayers at the scanning rate of 1000 mV/s, (f) Nyquist plot for the multilayer film with 12 bilayers in the frequency range from 4000 kHz to 100 mHz at an AC amplitude of 10 mV.

Coagulation self-assembled graphene macropore network

In this section, we develop a novel and easy-handling coagulation approach to construct 3D porous graphene network.[19] An aqueous graphene-oxide dispersion is a negatively charged colloidal system owing to the existence of ionizable groups, such as carboxyl and hydroxyl, on the surfaces of the nanosheets.[20] It has been reported that introduction of electrolytes, acids or polyelectrolytes will destroy the stability of a graphene-oxide colloid dispersion and cause the agglomeration of graphene nanosheets to form precipitates.[20, 21] Electrochemical double-layer capacitors (EDLCs), which are important for energy storage applications, use electrode materials with large surface area to store charges through electric double layer at the electrode/electrolyte interface.[20] The agglomerate graphene is a suitable candidate for EDLC applications, when the coagulation process is controlled to acquire desirable morphology, compactness, and inner structure. However, to our knowledge, few research efforts have been focused on the morphology and structure of the graphene coagulation assemblies to obtain high-performance EDLCs.

Macroporous graphene network was prepared through polyethylenimine(PEI)-induced coagulation of graphene-oxide nanosheets from dispersion (Figure 2). The supercapacitor made of graphene assemblies exhibited the high surface areal specific capacitance of 46.3 μF/cm^2, excellent rate performance, and long cyclic life, which is attributed to the face-to-edge aggregation of exfoliated single-layer graphene nanosheets (Figure 3).

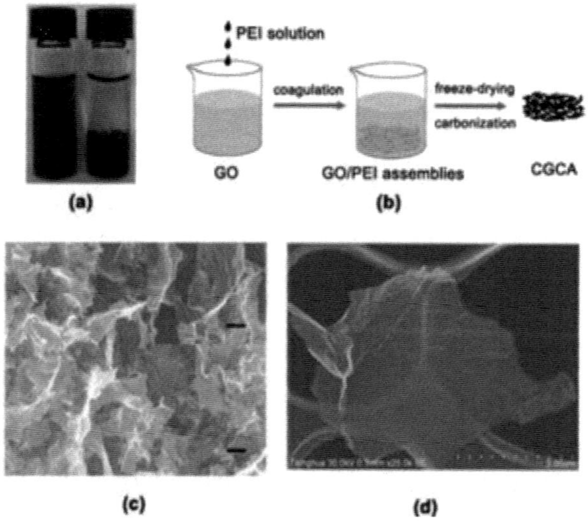

Figure 2. (a) Photographs of GO-water dispersion and GO/PEI coagulation. (b) Schematic representation of the prepared procedure of CGCA. (c) SEM images of the prepared CGCA. The inset shows a surface corrugated graphene sheet, the scale bar is 10 μm. (d) SEM images of exfoliated graphene layers from CGCA.

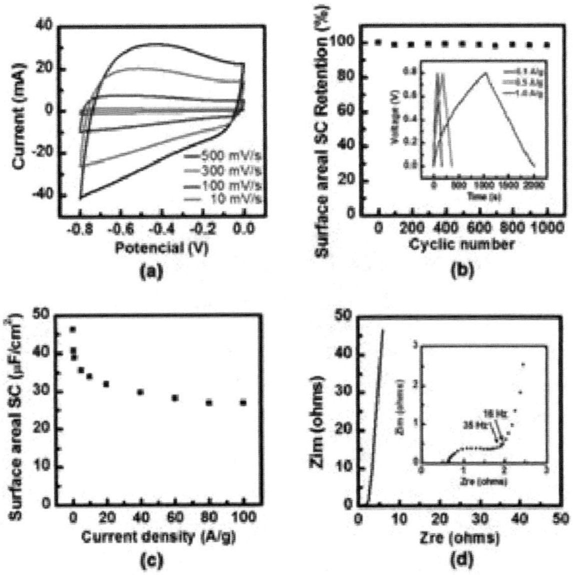

Fig. 3. (a) Cyclic voltammograms; (b)Cyclic performance and GCD curves; (c) Rate performance; (d) Nyquist plots for CGCA-based supercapacitor.

Hydrogel post-reduced hierarchical porous graphene films

In this section, we report a novel method developed to feasibly fabricate flexible large-area hierarchical porous graphene films by utilizing GO hydrogel as initial feedstock (Figure 3) [23]. The GO hydrogel films were prepared by convenient blade-casting technique and then reduced by hydroiodic acid/acetic acid (HI/HAc) under low temperature. For the first time, the current study demonstrates that in contrast to poor processibility of an aqueous GO dispersion, the viscous GO hydrogel is a highly processible material and can be readily processed to films with desired sizes and shapes. After fabricated to thin films with suitable sizes, the mechanically tough graphene electrodes with macroporous/mesoporous structures are feasibly obtained by the reduction of the stable GO hydrogel with HI/HAc (Figure 4). The flexible supercapacitor fabricated with the RGO electrodes exhibits high areal specific capacitance (71.0 mF cm^{-2} at 1 mA cm^{-2}), excellent rate performance (79.0 % retention at 100 mA cm^{-2}), and long cycle life (98.3 % retention after 5000 cycles). Astonishingly, the flexible supercapacitor prepared with the electrodes can be bent to arbitary angles (even 180°) with an almost negligible capacity loss (Figure 5).

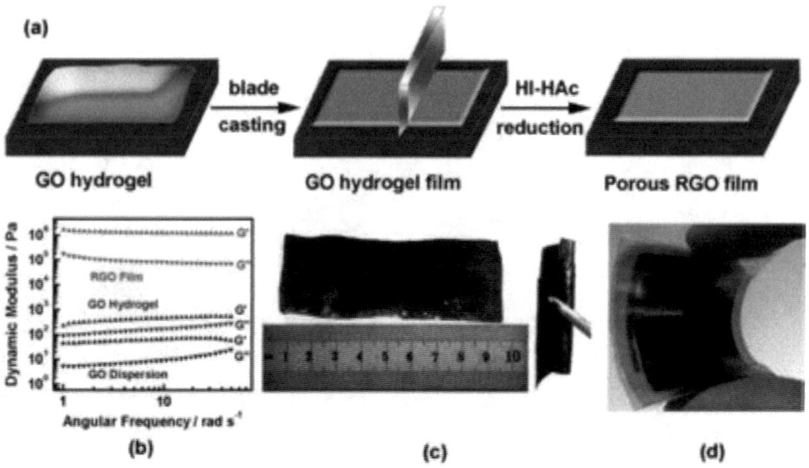

Figure 3. a) Schematic illustration of the fabrication process of the porous RGO film. b) Dynamic rheological behavior of the GO dispersion, GO hydrogel and RGO film. c, d) Photographs of the RGO film and the corresponding flexible supercapacitor. The GO disperion with the concentration of 6.8 mg mL-1 was used for the fabrication.

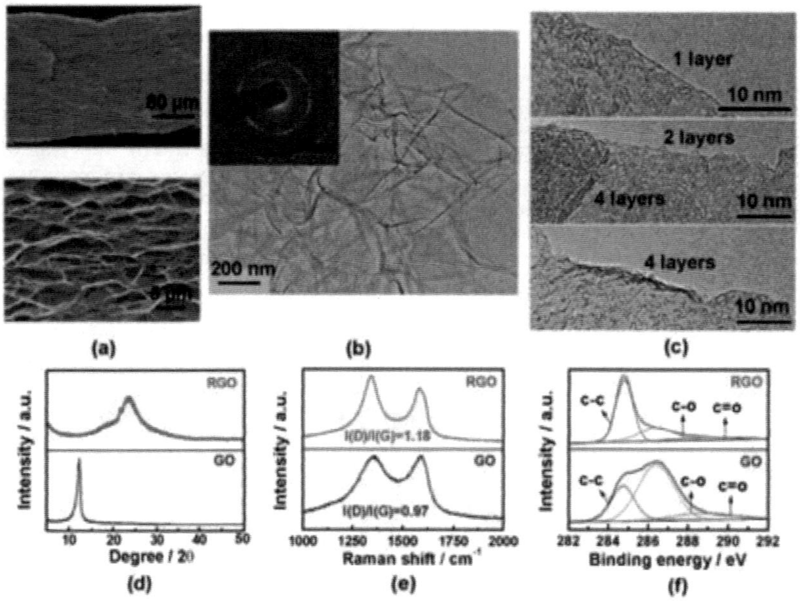

Figure 4. a) SEM images of the RGO films. b) TEM image and electron diffraction pattern of the RGO film. c) High resolution TEM images of the RGO films. d) XRD curves, e) Raman spectra, f) XPS spectra of the GO and RGO samples. The GO dispersion with the concentration of 6.8 mg mL^{-1} was used for the fabrication.

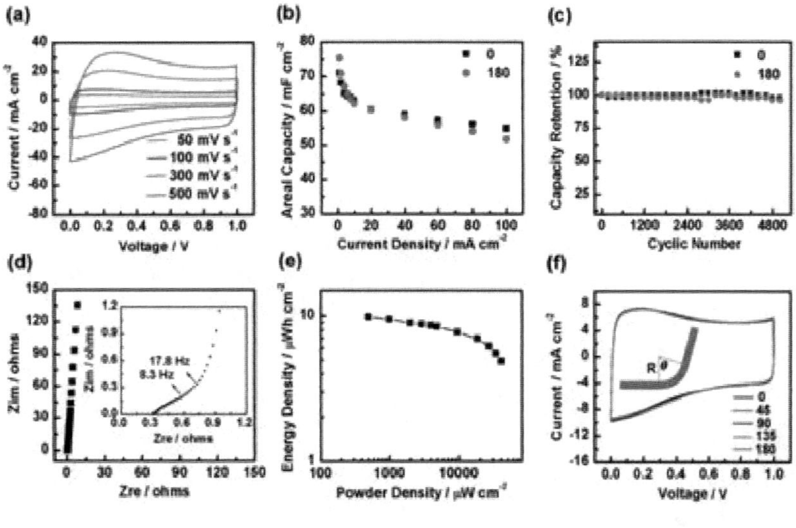

Figure 5. Performance of the film supercapacitor (1 mg cm^{-2}): a) CV curves under different scan rates in H_2SO_4 (1 M); b, c) Areal specific capacitance and cyclic life at the normal and 180° bending states; d) Nyquist plot; e) Ragone plot; f) CV curves of the film supercapacitor at the different bending states.

Conclusion

Fabricating unique structure with individual graphene layers to fully utilize the specific surface area of graphene is still a great challenge while many approaches have developed to construct various graphene architectures. In our paper, graphene has been effectively separated and exfoliated in the fabricated structure. Nowadays, chemically converted graphene can be produced at ton yield with cheap price, making this simple coagulation assembly approach possible for industrial applications. In addition, other faradaic pseudocapacitor materials, such as conducting polymers and various transition metal oxides, can be introduced as charged coagulants to further enhance energy and power density of the supercapacitor.

References

1. Novoselov, K. S.; Geim, A. K.; Morozov, S. V.; Jiang, D.; Zhang, Y.; Dubonos, S. V.; Grigorieva, I. V.; Firsov, A. A. *Science* **2004,** 306 (5596), 666-669.
2. Park, S.; Ruoff, R. S. *Nat. Nanotechnol.* **2009,** 4 (4), 217-224.
3. Li, D.; Muller, M. B.; Gilje, S.; Kaner, R. B.; Wallace, G. G. *Nat. Nanotechnol.* **2008**, 3 (2), 101-105.
4. Stankovich, S.; Dikin, D. A.; Dommett, G. H. B.; Kohlhaas, K. M.; Zimney, E. J.; Stach, E. A.; Piner, R. D. *Nature* **2006,** 442 (7100), 282-286.
5. Conway, B. E. *Electrochemical Supercapacitors: Scientific Fundamentals and Technological Applications*; Kluwer Academic/Plenum Publisher: New York, **1999**.
6. Winter, M.; Brodd, R. J. What are batteries, fuel cells, and supercapacitors? *Chem. Rev.* **2004**, 104 (3), 4245-4270.
7. Simon, P.;Gogotsi, Y. Materials for electrochemical capacitors. *Nat. Mater.* **2008**, 7 (11), 845-854.

8. Miller, J. R.; Outlaw, R. A.; Holloway, B. C. *Science* **2010**, 329 (5999), 1637-1639.
9. Zhu, Y.; Murali, S.; Stoller, M. D.; Ganesh, K. J.; Cai, W.; Ferreira, P. J.; Pirkle, A.; Wallace, R. M.; Cychosz, K. A.; Thommes, M.; Su, D.; Stach, E. A.; Ruoff, R. S. *Science* **2011**, 332 (6037), 1537-1541.
10. El-Kady, M. F.; Strong, V.; Dubin, S.; Kaner, R. B. *Science* **2012**, 335 (6074), 1326-1330.
11. Segal, M. *Nat. Nanotechnol.* **2009**, 4 (10), 612-614.
12. Xia, J.; Chen, F.; Li, J.; Tao, **2009**, 4 (8), 505-509.
13. Zhang, K.; Zhang, L. L.; Zhao, X. S.; Wu, J. *Chem. Mater.* **2010**, 22 (4), 1392-1401.
14. Si, Y.; Samulski, E. T. *Chem. Mater.* **2008**, 20 (21), 6792-6797.
15. Decher, G. *Science* **1997**, 277 (5330), 1232-1237.
16. Srivastava, S.; Kotov, N. A. *Acc. Chem. Res.* **2008**, 41 (12), 1831-1841.
17. Mamedov, A. A.; Kotov, N. A.; Prato, M.; Guldi, D. M.; Wicksted, J. P.; Hirsch, A. *Nat. Mater.* 2002, 1 (3), 190-194.
18. Z. Y. Xiong, T. H. Gu and X. Wang, Langmuir, 2014, 30, 522-532.
19. Z. Y. Xiong, C. L. Liao, X. G. Wang, J. Mater. Chem. A 2014, 2, 19141.
20. D. Li, M. B. Muller, S. Gilje, R. B. Kaner and G. G. Wallace, Nature nanotechnology, ***2008***, 3, 101-105.
21. Y. Tian, Y. W. Cao, Y. Wang, W. L. Yang and J. C. Feng, Advanced Materials, ***2013***, 25, 2980-2983.
22. M. F. El-Kady, V. Strong, S. Dubin and R. B. Kaner, Science, ***2012***, 335, 1326-1330.
23. Z. Y. Xiong, C. L. Liao, X. G. Wang, Adv. Mater. 2015,27, 4469.

426077 N-Doped Material for Supercapacitor from Lignin Biomass

Abstract:

N-doped material for supercapacitor from lignin biomass

Muslum Demir[1], Ahmed A. Farghaly[2], Burak Aksoy[3], Timur Islamoglu[2], Harry T. Cullinan[3], Hani M. El-Kaderi[2], Maryanne M Collinson[2] and Ram B. Gupta[1,]

1) Department of Chemical and Life Science Engineering, Virginia Commonwealth University, Richmond, VA 23284

2) Department of Chemistry, Virginia Commonwealth University, Richmond, VA 23284

3) Alabama Center for Paper and Bioresource Engineering, Dept. of Chemical Engineering, Auburn University, Auburn, AL 36849Abstract

Abstract

Lignin is a high volume byproduct produced from the pulp and paper industry. Most of lignin is currently burned to generate electricity. The industry has been searching for high value added usage of lignin to improve the process economics. . On the other hand, sustainable energy supplies depends on the progress of green technologies needs to be developed urgently which can be achieved via environmentally friendly approaches with inexpensive precursors and rich resources obtained from bio-mass. These materials has recently been employed in the field of energy such as fuel cells and supercapacitors In this work, lignin biomass which is a cross-linked polymer of phenols is converted into N-doped porous carbon using a two-step conversion process. Lignin is first converted into biochar via hydrothermal carbonization at 300 °C and 1500 psi using very unique reactor which is capable of applying sonication to biomass at high T,P. Then, N-doped porous carbons were obtained by mixing biochar with a nitrogen rich source and $ZnCl_2$ at high temperture (700-1000 °C) and under an inert atmosphere at 15 psi. N-doped porous carbons effectiveness of three different temperature is examined like morphology, thermal stability, surface properties. The morphology, thermal stability, and surface properties of the as-prepared N-doped porous carbons at different temperatures (700, 850 and 1000 °C) were examined. The capacitive characteristics of the as-prepared materials are investigated with cyclic voltammetry (CV) at 1 mV to 500 mV. N-doped porous carbon prepared at 850 °C demonstrated the highest capacitance among the different samples (280 F/g at 1 mV CV).

426615 Nanostructured Surfactant Ionic Liquids with Unusually High Capacitances for High-Temperature Flexible Supercapacitors

Abstract:

We report a new class of electrolytes, surfactant ionic liquids (SILs), that give rise to unprecedentedly high intrinsic areal capacitances (~100 to 200 μF/cm^2), much larger than the values (~5 to 20 μF/cm^2) of other known ionic liquids (ILs) based on smaller anions as well as commonly used aqueous/organic electrolytes. Such high capacitances can probably be attributed to an increased molecular structuring at the nanometer scale due to the presence of large non-polar alkyl chains and charged heads in these fluids. We demonstrate that extremely high gravimetric/volumetric capacitances can be achieved by combining SILs with proper porous electrodes with high surface areas. Due to the extremely wide potential working windows of ILs (~ 4 to 5 V), we show that the SIL-based devices exhibit ultrahigh energy densities. In addition, the SIL-based devices show an increasing capacitance with higher temperatures, which highlights another unique advantage of SILs, which is the suitability for operation at extremely high temperature (~200 ^{o}C), with myriad niche applications such as powering sensors and actuators for oil industry downhole operations (above 120 ^{o}C), electronics in hybrid electric vehicles (HEV, >60 ^{o}C), or military weapons and space equipment. Furthermore, we demonstrate that SILs can be easily polymerized to generate a quasi-solid yet flexible gel-like structure, as well as form seamless binding with current collectors/porous electrodes, for the development of next-generation solid-state highly flexible energy devices.

428157 Spinel Decorated Aligned Carbon Nanotube Arrays As Supercapacitor Electrodes

Abstract:

Accompanying the need for green energy generation is a need for energy storage systems since most green energy sources like solar and wind are intermittent. In recent times, much research focus has been placed on providing more durable and higher capacity energy storage devices like supercapacitors which has led to the design of better electrode materials for these supercapacitors. Graphitic carbon nanomaterials have been tipped as superior materials for making supercapacitors because of their unique electrical properties and pore structure. We report the use of aligned carbon nanotubes as supercapacitor electrodes and highlight their superiority, as a result of a better ordered pore structure, to entangled or randomly oriented carbon nanotubes for electrode applications. Also, the enhanced capacitance of these electrodes after electrochemical decoration with transition metal oxides, specifically spinels, is studied and reported. Specifically, we use Ni-Co mixed oxide and Cu-Mn mixed oxide spinels characterized using EDX and XRD. These spinels have been chosen because of their relatively high electrical conductivity and are optimized for relative metal ratios. Different methods of electrochemical deposition are also explored to achieve uniform coating. Moreover, we reveal a new universally applicable method to preserve pore structure and prevent forest collapse after electrodeposition on carbon nanotube forests that are not free-standing (i.e. not grown with PECVD).

434817 Thin Film Manganese Oxide Electrodes for Batteries and Supercapacitors As Grown By Atomic Layer Deposition

Abstract:

Manganese dioxides are commonly used as electrode materials for charge storage in supercapacitors (*e.g.* α-MnO_2) and cation intercalation batteries (*e.g.* λ-MnO_2). Manganese-based electrode materials have particular appeal in these applications due to the earth abundance and low cost of manganese. Nanoscale thin films are expected to increase the specific capacity and rate capability of manganese oxide electrodes by providing shorter bulk diffusion lengths for cations. Here we demonstrate that nanoscale thin films of MnO grown by atomic layer deposition (ALD) using bis(ethylcyclopentadienyl)manganese ($Mn(CpEt)_2$) and water can be electrochemically oxidized in aqueous electrolytes to produce electrochemically active manganese dioxide for batteries and supercapacitors. However, our results show that the electrochemical oxidation of MnO ALD films results in partial dissolution and delamination. To avoid this undesirable material loss, we have developed an approach to directly grow more oxidized manganese oxide films by ALD using ozone as the oxidant and $Mn(CpEt)_2$ as the Mn precursor. We have also used intermediate ozone doses during $Mn(CpEt)_2$ and water exposures during ALD growth. The use of ozone results in more oxidized manganese oxide films without the electrochemical oxidation step.

432788 First-Principles-Based Analysis of the Potential Benefits of Graphene Oxide for Supercapacitor Applications

Abstract:

In recent years, graphene oxide has been widely-explored as an alternative electrode material for use in supercapacitors. One of the primary drivers for interest in this material is their ease in solution-based processing, which is highly scalable. Experimental studies have demonstrated encouraging performance through utilization of these materials, although the observed benefits appear widely scattered and sensitive to the reduction process. To help clarify the influence of graphene oxide reduction on the interfacial capacitance, we perform combined density functional theory and molecular dynamics simulations to study the influence of oxygen-moiety coverage on graphene at different O:C ratios. In this talk, we present the sensitivity of the interfacial capacitance to O:C by decoupling the contributions from the quantum and electric double layer capacitances. In particular, we demonstrate that, under certain conditions, the overall capacitance can be most enhanced when the quantum capacitance effectively mitigates the double layer capacitance. Our findings clearly demonstrate that graphene oxide is a promising material to consider, in which careful control and optimization of both the concentration and composition of oxygen-containing groups along the basal surface will be required to achieve superior performance in supercapacitors.

427061 A Self-Supported and Long-Life Li-Se Battery Cathode Enabled By 3D Mesoporous Carbon/Graphene Hierarchical Architecture

Abstract:

For 2015 AIChE Annual Meeting

Topical 5: Nanomaterials for Energy Applications

Session: Nanomaterials for Energy Storage

A Self-supported and Long-life Li-Se Battery Cathode Enabled by 3D Mesoporous Carbon/Graphene Hierarchical Architecture

Kai, Han, Central South University (China)* hankai@csu.edu.cn

Heng, Jiang, Central South University (China) *Hongqi, Ye, Central South University (China)*

Zhao, Liu, Northwestern University (USA) *Fang Dai, GM, Global R&D (USA)*

Lithium-ion batteries have been widely used as one of the most important energy storage techniques in the past decades. However, the energy density of current Li-ion batteries is insufficient to satisfy the requirements for emerging technologies, such as electric vehicles and smart grid, which is mainly limited by the low capacity of cathode materials. Therefore, developing high-capacity cathode materials are urgently needed. Although sulfur could offer high theoretical lithium storage capacity when applied as cathode, the Li-Sulfur batteries significantly suffer from poor cycling stability due to the insulating nature of sulfur and shuttling effect of polysulfides. Most recently, selenium from same group as sulfur has been considered to be a promising alternative for high energy and long cycle life Li batteries due to its comparable volumetric capacity to but much better electrical conductivity than sulfur. However, Li-Se system also faces two similar issues: shuttling of polyselenides and volume change during charge/discharge processes.

In this work, we strategically design and fabricate a self-supported selenium composite with a 3D hierarchical architecture constructed by mesoporous carbon and graphene, in which selenium is impregnated into the mesoporous carbon nanoparticles (Se/MCN), followed by further embedding between graphene sheets (Se/MCN-RGO). Structural characterization by XRD and SEM/TEM provide clear evidence for the unique structure. Such architecture not only provides the electrode with excellent electrical and ionic conductivity, but also efficiently suppresses polyselenides shuttling and accommodates volume change during charge-discharge cycling. At selenium content of 62%, the Se/MCN-RGO cathode exhibits not only high discharge capacity of 655 mAh g-1 at 0.1 C (97% of theoretical capacity) but also long cycling stability under high rate (a very small capacity decay of 0.008% per cycle over 1300 cycles at 1C). It is furthermore believed that the concept of such 3D hierarchical carbon architecture could be easily expanded to other advanced energy storage systems, such as silicon for Li-ion batteries and metal oxide for supercapacitors. The structural and electrochemical characterization results of the high-performance Se/MCN-RGO composite will be presented.

Keywords: Long-life Li-Se batteries; selenium cathode; mesoporous carbon; graphene

432469 Synergy Between Different Graphene Structures in Layer-By-Layer Assembly for Li-Air Battery Cathodes

Abstract:

Lithium-air or lithium-oxygen batteries widely known for its high energy density stand in the spotlight of world with an expectation that it could promote electric vehicles or high grid energy storages to the next level. Studies on the interface between carbon and catalyst or the reaction properties of each with respect to oxygen and electrolyte are underway. However, a few was reported on the structural characteristic of a cathode coated with two different carbons without a catalyst. In the current study, a sandwich-structured cathode was manufactured by successively alternating two different carbons in a layer-by-layer assembly manner onto a gas diffusion substrate. With a constant amount of total carbon but with a different number of layers a clear synergic effect was demonstrated in terms of both increasing the specific capacity and reducing charge overpotential. The results strongly indicate that the structural variance of cathode controlled by different carbon morphologies affects the overall performance of lithium-air batteries, not to mention their surface properties.

Characterization of Graphene-Metal/Metal Oxide Composites

Donald Johnson
Tuskegee University
Graduate Student
November 10, 2015

TUSKEGEE UNIVERSITY

15AIChE
Annual Meeting, Salt Lake City, UT

Presentation Outline

- Background
- Specific Objectives of the research
- Experimental Method
- Characterization (SEM,TOC,XRD)
- Electrochemical Characterization (CV)
- Acknowledgements

Background

- With the growing importance of energy applications, a need has emerged to characterize the various forms of activated and nanostructured carbons in the context of their electrochemical charge-storage capacity.
- Graphene has become a promising material due to its unique structure and outstanding intrinsic physical properties, such as high electrical conductivity, strong mechanical properties, and large surface area
- Three-dimensional (3D) graphene aerogel (GA) has recently attracted much attention because of the combination of a 3D porous structures and excellent intrinsic properties of graphene

Background Cont.

- Graphene Aerogel has great versatility both at the nanoscopic level and the macroscopic level.
- Pore size distribution, Pore volume, and surface area are all tunable properties at the synthesis level which produces a wide spectrum of unique material properties.
- Compared t0 2D graphene sheets, 3D graphene aerogel allows for gas to flow easily through the structure, and enables easier control of reaction characteristics based on vessel allowing access to large scale applications.

Applications of Graphene-Metal/Metal Oxide Composites

- Two major areas of applications:
  - Electrode materials in electrochemical energy storage systems (e.g. as electrodes in supercapacitors and lithium ion batteries.
    - Produce alternative high capacity energy storage systems with long life and high cyclibility
  - Electrocatalysts in the electrochemical reduction of carbon dioxide (CO_2) into environmentally friendly and usable products
    - Reducing atmospheric CO_2 reduces global warming

Specific Objective

- To synthesize graphene/metal and graphene/metal oxide composites
- Use Electrochemical methods to characterize the composites materials
  - Why these composite materials
    - Take advantage of the electrocatalytic properties of metallic (e.g. Pt and Ni) and metal oxide particles (e.g. Co_3O_4 & MnO_2) as well as the high surface area of graphene to create materials of superior electrochemical properties
  - To achieve:
    - Develop an improved structured electrocatalyst using bottom-up synthesis method
    - Perform electrochemical characterization of the prepared electrocatalyst

Experimental Methods

- Two Categories
  - Electrode Materials (Catalyst) Development
    - Materials Synthesis
      - Graphene Aerogel supported Electrode
  - Electrochemical Characterization

Graphene Aerogel Support

- Electrode Material Synthesis:
  - Platinum and Nickel powder (150 -400 nm)
  - Manganese Oxide
- Deposition Method:
  - Simultaneous Deposition
  - In situ Deposition

Schematic of Materials Synthesis

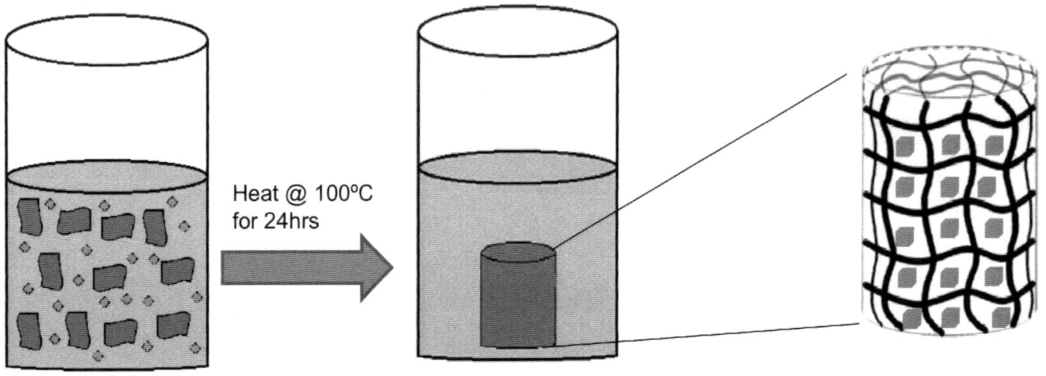

Graphene Oxide solution with metal nanoparticles

Graphene Aerogel with metal particles

Interior schematic of Graphene Aerogel with metal particles

Experimental Procedure

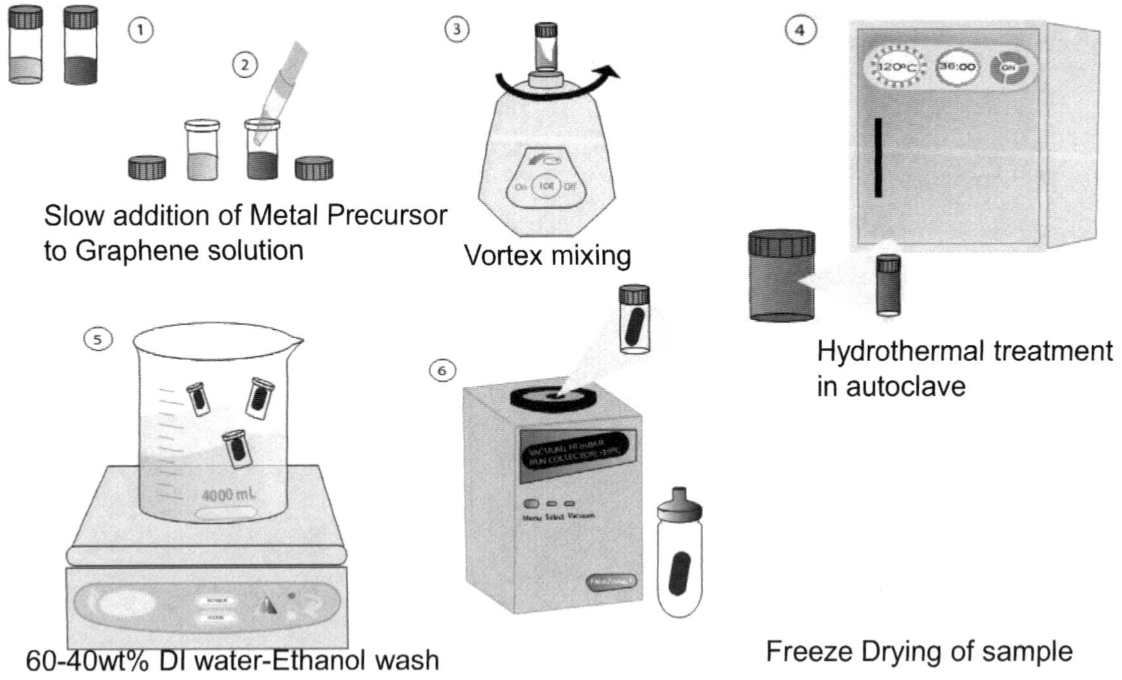

Graphene Oxide to Graphene Aerogel

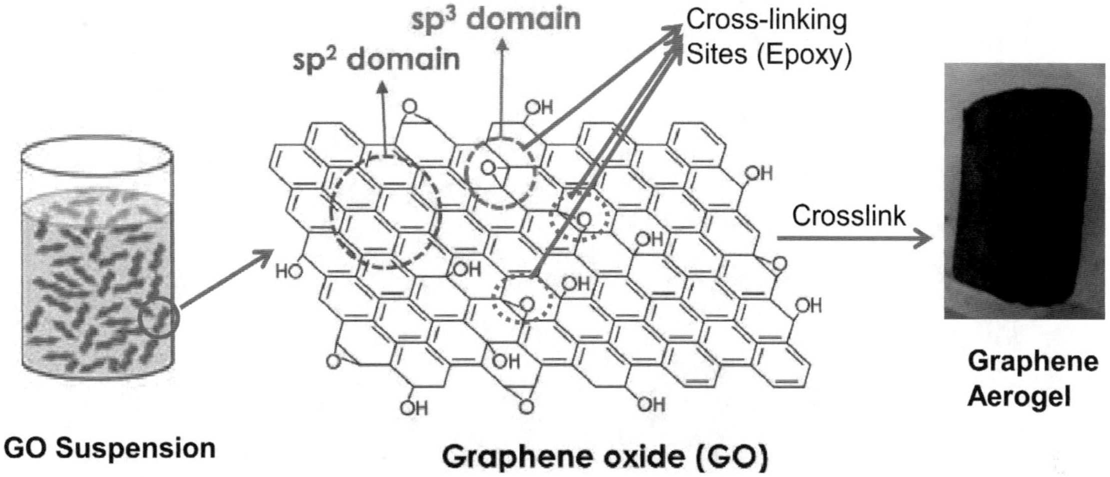

Why Freeze Drying ??

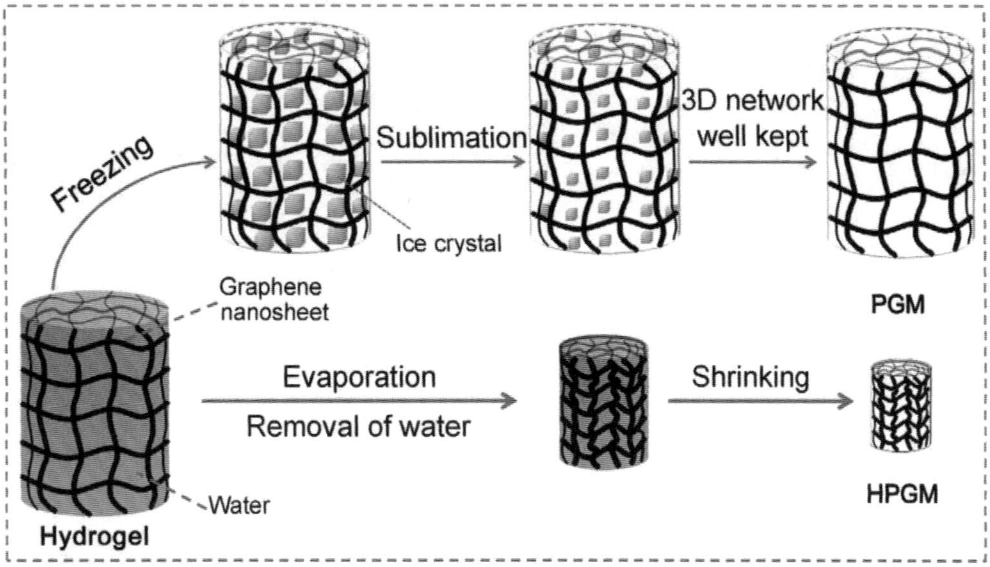

- Zhao G, Wen T, Chen C, Wang X. Synthesis of graphene-based nanomaterials and
their application in energy related and environmental-related areas.

Graphene Aerogel w/o particles

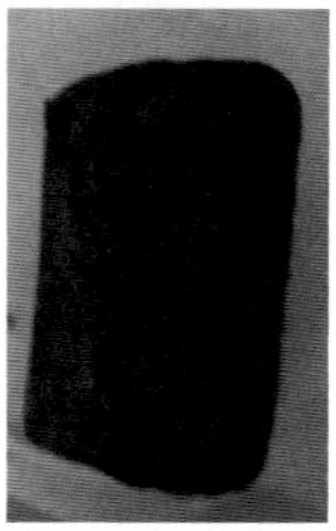

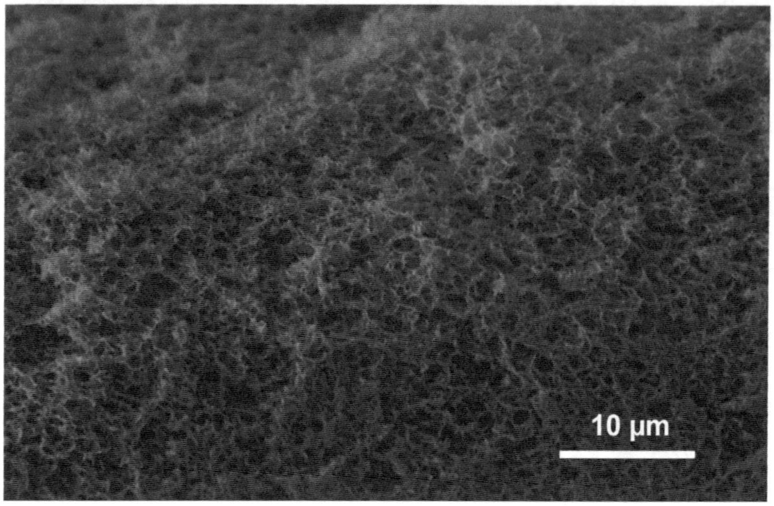

Bulk Graphene Aerogel

Cross-sectional SEM of Graphene Aerogel

Graphene Aerogel-Ni composite

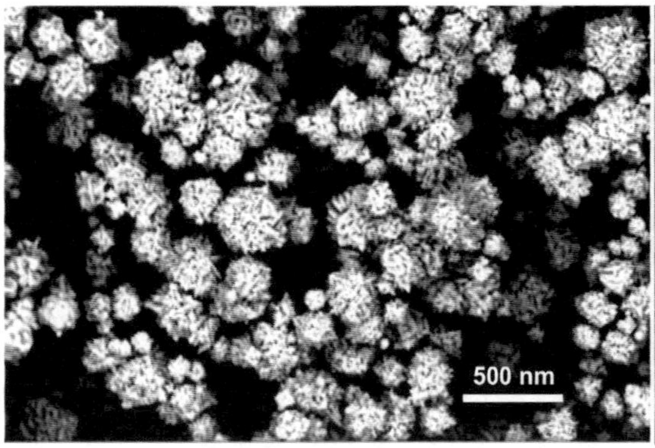

SEM of nickel particles

Cross-sectional SEM of Graphene Aerogel w/ Nickel particles

Graphene Aerogel-Pt Composite

SEM of Platinum particles

Cross-sectional SEM of Graphene Aerogel
w/ Platinum particles

Graphene Aerogel-MnO Composite

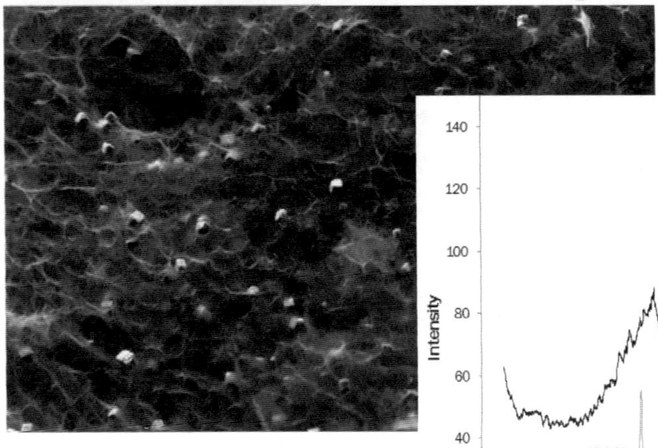

Cross-sectional SEM of Graphene Aerogel w/ MnO particles

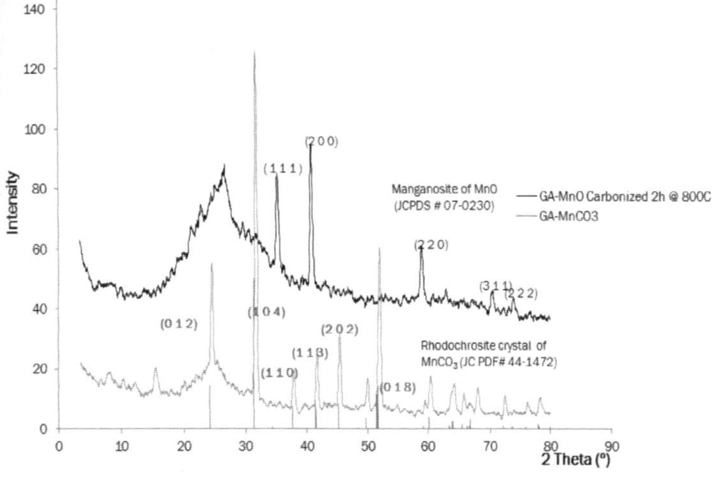

XRD of graphene MnO composite

Rigaku D/Max 2200 X-Ray Cu kα radiation, λ=1.54Å, 40kV & 30mA

Aerogel Total Organic Carbon (TOC)

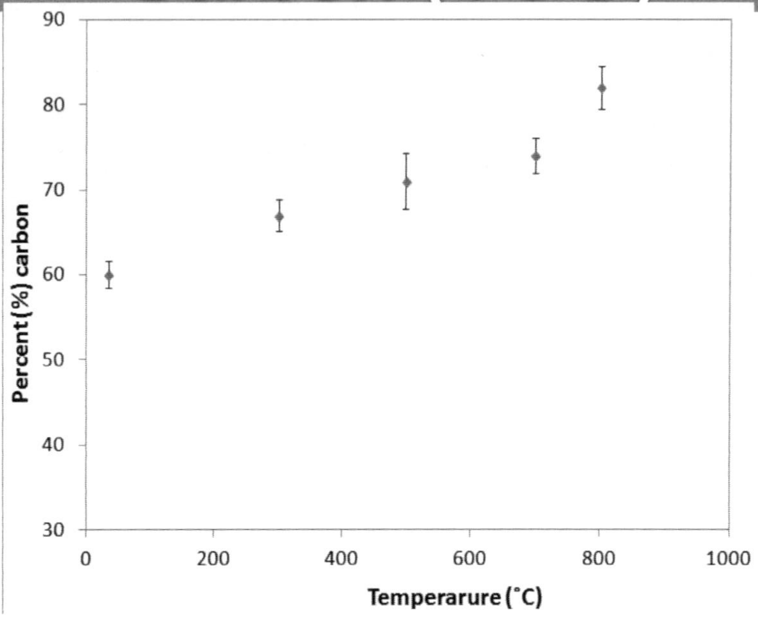

- The carbon content of materials is enrich with increasing carbonization temperature

Cyclic Voltammetry (CV)

- An electrochemical technique which measures the current that develops in an electrochemical cell where voltage is applied.
- CV is performed by applying the potential between the working electrode and the reference electrode while the current is measured between the working electrode and the counter electrode.
- Anodic and Cathodic potential can be determined from the Voltammogram

CV Experimental Setup

- The following experimental setup for methanol oxidation was performed to obtain a cathodic/anodic voltammogram:

- 0.5M CH_3OH oxidation in 0.5M H_2SO_4

  - Performed after 30 mins N_2 gas bubbling

- Specs: Gamry, Standard Dr. Bob's Cell, 50mL

CV: GA Methanol Oxidation

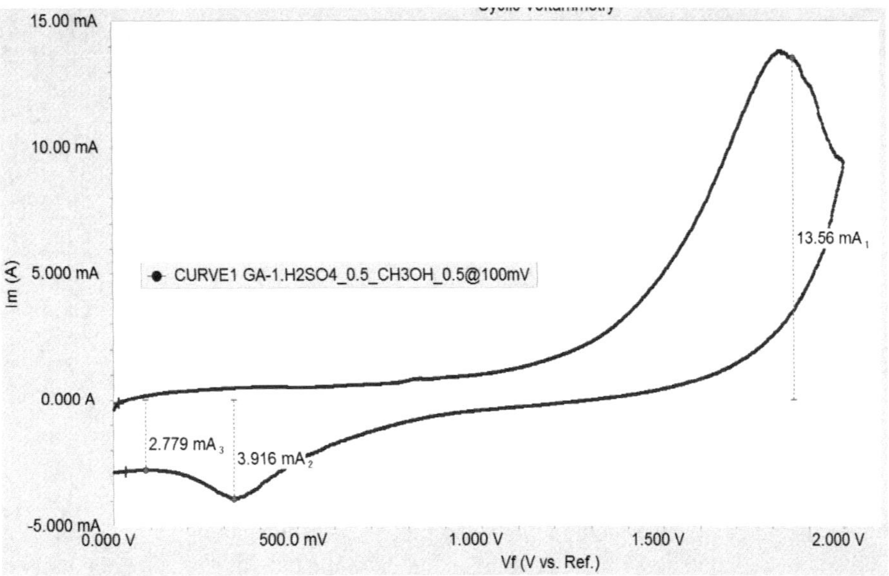

Cyclic Voltammogram of Graphene Aerogel in methanol oxidation with 100mV scan rate

CV: Pt wire v. GA Methanol Oxidation

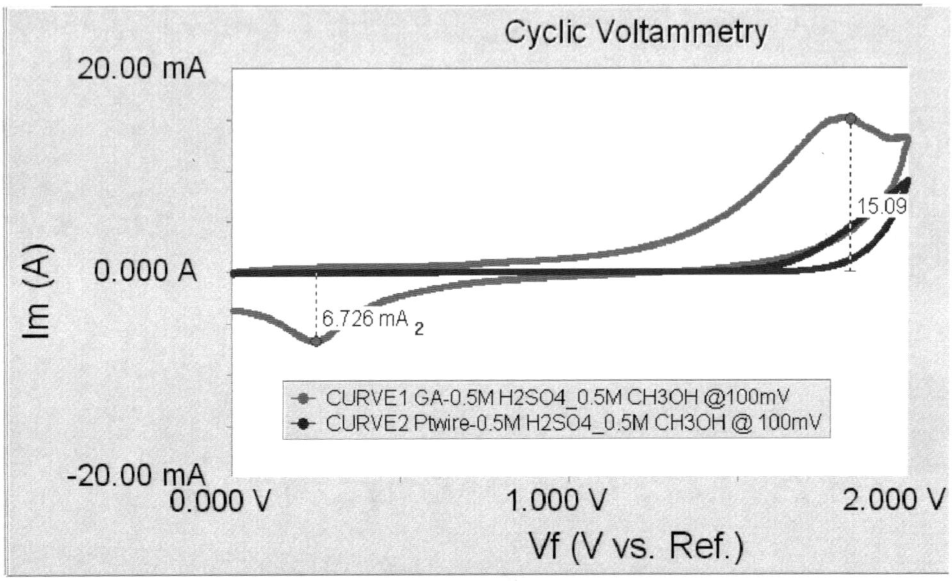

Cyclic Voltammogram of Platinum wire vs. Graphene Aerogel in methanol oxidation with 100mV scan rate

CV: GA-Pt Methanol Oxidation

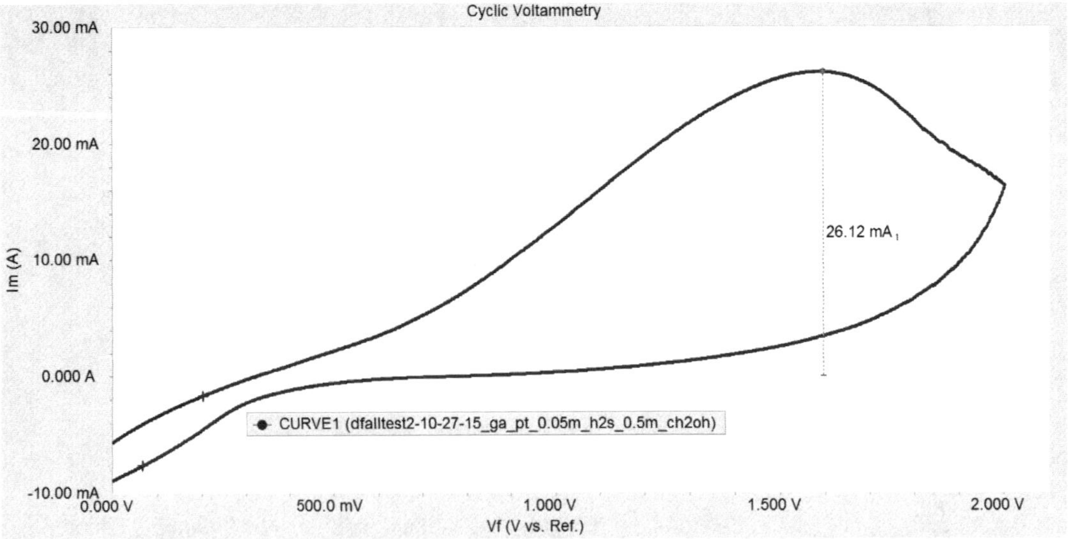

Voltammogram of Graphene Aerogel-Platinum composite in Methanol oxidation,
100mV scan rate

CV: GA-Ni Aerogel in CH_3OH Oxidation

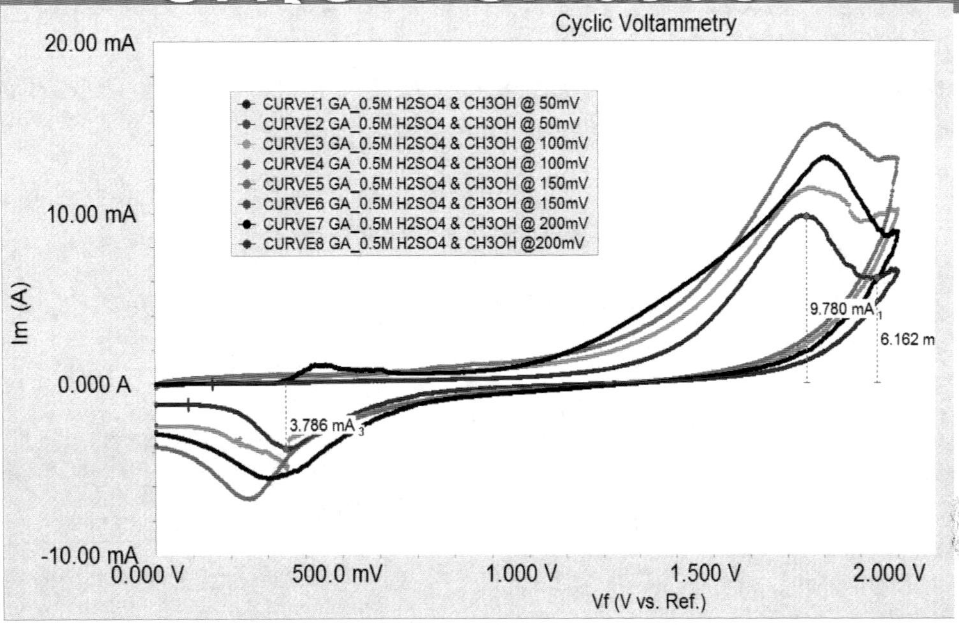

Overlay Voltammogram of Graphene-Nickel composite in methanol oxidation

CV: GA-MNO Methanol Oxidation

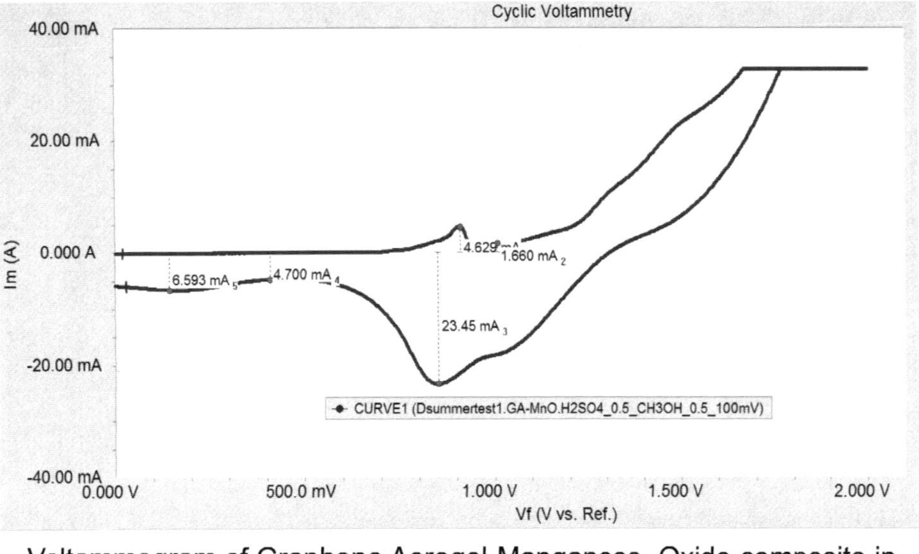

Voltammogram of Graphene Aerogel-Manganese Oxide composite in Methanol oxidation with100mV scan rate

Selected Reference

- Qinghong Huang et al. One-step Synthesis of Pt Nanoparticles Highly Loaded on Graphene Aerogel as Durable Oxygen Reduction Electrocatalyst 2015
- Zhao G, Wen T, Chen C, Wang X. Synthesis of graphene-based nanomaterials and their application in energy related and environmental-related areas.
- Zhang N, Zhang Y, Xu YJ. Recent progress on graphene-based photocatalysts: current status and future perspective nanoscale 2012;4:5792–813.
- Shao Y, Wang J, Wu H, Liu J, Aksay IA, Lin Y. Graphene based electrochemical sensors and biosensors: a review electroanalysis 2010;22:1027–36.
- Choi HJ, Jung SM, Seo JM, Chang DW, Dai L, Baek JB. Graphene for energy conversion and storage in fuel cells and supercapacitors. Nano Energy 2012;1:534–51.

Acknowledgement

- Special Thanks to the funding agency:
  - DOE-NNSA
- Also a thanks to my Dept. Head, Dr. Vahdat
  - Other Faculty:
    - Dr. Mbah,
    - Dr. Baah,
    - Dr. Begum,

417072 Co-Continuous Composite Electrodes Derived from Bijels That Concurrently Deliver High Energy and Power

Abstract:

Pressing energy demands for national security, distributed power, and portable technologies require the development of energy storage devices that can simultaneously offer high power and energy densities. In order to achieve this goal, the kinetics of ion and electron transport within the electrodes must be enhanced while maintaining a large capacity for energy storage. Therefore, the idealized electrode microstructure has been envisioned as a three-dimensional, co-continuous arrangement of percolating passages for efficient ion and electron transport that would also enable a large volumetric density of active material. Herein, we report a novel method to produce this distinctive microstructure through the use of soft matter templates derived from bicontinuous interfacially jammed emulsion gels (bijels). These soft materials inherently have a co-continuous microstructure, where two interpenetrating fluid domains are kinetically arrested through jamming of particles at the fluid-fluid interface during spinodal decomposition. Bijels can be converted to electrolytically active composites through simple chemical post processing steps. Our synthesis procedure allows for independent tuning of the microstructural parameters that govern the electrochemical performance of these electrodes. In this talk, we present composite electrodes that we have synthesized by this technique and demonstrate how their electrochemical characteristics are dictated by these morphological parameters, allowing for concurrent delivery of high energy and power, bridging the gap between modern batteries and supercapacitors.

Hierarchical Carbon-based Electrode Materials for Vanadium Redox Flow Batteries

P. Trogadas, T.P. Neville, D.J.L. Brett, P.R. Shearing, and M.-O. Coppens

Centre for Nature Inspired Engineering and Electrochemical Innovation Lab, Department of Chemical Engineering, University College London, Torrington Place London WC1E 7JE United Kingdom.

The demand for efficient energy storage has necessitated the search for novel electrode materials for energy storage devices, such as the vanadium redox flow battery (VRFB).[1] To maximize storage capacity and kinetics during charge and discharge cycles, electrode materials must have large surface area, high electronic and ionic conductivity as well as long-term stability. Hierarchical nanostructured electrode materials have attracted much attention nowadays due to their unique properties compared to bulk materials; the interconnectivity between pores of different sizes and the high surface area results in improved diffusivity and utilization of reactants and thus improved battery performance. Key geometric characteristics of meso / nanoporous carbon based materials will be presented along with their synthesis technique and activity / stability measurements. Using a combination of micro-tomography (CT) measurements of voltage-cycled graphite felts (over ca. 30 charge / discharge cycles), scanning electron microscopy (SEM), and X-ray photoelectron spectroscopy (XPS) measurements of the same felt electrode, the first description of micro-structural evolution processes occurring in VRFB felts during operation is provided.

References

1. A. Weber et al., J. Appl. Electrochem., 41, 1137 (2011).

415031 Artificial Photosystem I and II: Highly Selective Solar Fuels and Tandem Photocatalysis

Abstract:

Artificial photosynthesis, or generation of solar fuels from carbon-dioxide and water, can provide an important alternative for rising carbon-dioxide emission and renewable energy generation. The photosynthesis process performed by plants in an energetically frugal, albeit inefficient process, provides important design principles for success: separation of Photosystem II (PS680) and I (PS700) process, focus on selectivity of product over high efficiency (to save energy in separations), and optimizing the light-harvesting strategies. Here we show novel composite photocatalysts made from widebandgap nanotubes (artificial photosystem II) and different QDs (artificial photosystem I) to separate photo-oxidation and photo-reduction processes and improve material stability. By tuning the redox potentials using the size, composition and energy band alignment of QDs, we demonstrate highly selective (>greater 90%) and efficient production of ethane, ethanol and acetaldehyde as solar fuels, using different wavelengths of light. We also show that this selectivity is a result of precise energy band alignments (using cationic/anionic doping of nanotubes, QD size etc.), confirmed using measurements of electronic density of states. Furthermore, using hot-carriers in these quantum-confined materials, the selective fuels were changed to higher redox potential solar-fuels, using higher energy photons in a novel hot-carrier photocatalysis process. This novel color-selective production of specific solar fuels provides a new pathway to utilize broad-band solar radiation efficiently, improve selective generation of inexpensive solar fuels (like alkanes, alcohols, aldehydes and hydrogen), and all design of tandem structures (red, green, blue) from stacked selective catalysts, allowing almost full visible spectrum (410 730nm) utilization with different desired fuels produced simultaneously.

431413 Charge-Storage Mechanisms for High Surface Area Carbides and Nitrides

Tuesday, November 10, 2015: 5:15 PM
251F (Salt Palace Convention Center)

Abdoulaye Djire[1], Jason Siegel[2], Lilin He[3], Alice E. S. Sleightholme[4], Saemin Choi[1], Paul G Rasmussen[4] and Levi T. Thompson[4], (1)Chemical Engineering, University of Michigan, Ann Arbor, MI, (2)Mechanical Engineering, University of Michigan, Ann Arbor, MI, (3)Neutron Sciences and Scattering Division, Oak RIdge National Laboratory, Oak Ridge, TN, (4)Department of Chemical Engineering, University of Michigan, Ann Arbor, MI

Charge-Storage Mechanisms for High Surface Area Carbides and Nitrides

Abdoulaye Djire[a], Jason B. Siegel[b], Lilin He[c], Alice E. S. Sleightholme[a], Saemin Choi[ad], Paul Rasmussen[ad], and Levi T. Thompson[abd]*

[a]Department of Chemical Engineering

[b]Department of Mechanical Engineering

[c]Oak Ridge National Laboratory

[d]Hydrogen Energy Technology Laboratory

University of Michigan, Ann Arbor, MI 48109-2136, USA

Tel. (734) 936-2015

E-mail: ltt@umich.edu

Introduction

Early transition-metal carbides and nitrides are being considered for use as electrode materials in supercapacitors due to their high accessible surface areas, high electronic conductivities and low cost. They possess high pseudocapacitances, good capacitance retention during cycling and wide voltage windows. The nitrides of vanadium (VN) and molybdenum (γ-Mo_2N) have received the most attention due to their high pseudocapacitances [1]. Previously, we demonstrated, using ion isolation experiments, that H^+ and OH^- were the active ions that gave rise to the pseudocapacitance [2]. Using X-ray absorption spectroscopy to track changes in the oxidation state of V and Mo in the VN and γ-Mo_2N materials, respectively, a pseudocapacitive charge storage mechanism was proposed for both materials in aqueous electrolytes [3]. The distribution of H^+ and OH^- with applied potential (location and depth of storage), however, remains poorly understood. This poses significant challenges in the design and full exploitation of carbide and nitride materials for supercapacitors. Here we report a detailed investigation of the charge storage mechanisms for early transition-metal carbides and nitrides in aqueous media. The pseudocapacitive charge storage mechanism has been investigated using x-ray absorption spectroscopy and neutron scattering and a combination of electrochemical techniques including cyclic voltammetry (CV), and electrochemical impedance spectroscopy (EIS). The amount of inserted hydrogen and hydroxide per mole of electron during pseudocapacitive storage was determined, and pseudocapacitive charge storage mechanisms were proposed and compared to those previously reported [3]. Additionally, the location of inserted hydrogen and hydroxide, depth of storage, and distribution in pores were determined.

Experimental

High-surface-area carbides and nitrides of Ti, V, Nb, Mo, and W were prepared from TiO_2 (99.9%, Alfa Aesar), V_2O_5 (Alfa Aesar), Nb_2O_5 (99.9985%, Alfa Aesar), $(NH_4)_6Mo_7O_{24}.4H_2O$ (81-83% as MoO_3, Alfa Aesar) and WO_3 (Alfa Aesar) by temperature-programmed reaction (TPR) method with 15% CH_4 in H_2 (Cryogenic Gases) or NH_3 (Cryogenic Gases), respectively. Characterization of the structural properties was performed using nitrogen physisorption and X-ray diffraction. The total capacitance and extent of pseudocapacitance was determined based on results from CV and EIS. For selected materials, details regarding the insertion and location of H^+ and OH^- and metal oxidation state changes during electrochemical cycling were determined using neutron scattering and x-ray absorption.

Results and Discussion

Figures 1a and 1b show ratios of H^+/e^- and OH^-/e^-, for γ-Mo_2N and VN, respectively, as function of applied potential. There was insertion of $1H^+$ per $2e^-$ and removal of $2OH^-$ per $1e^-$ for these materials, as they were electrochemically cycled, suggesting reduction of the Mo and V metals. Reduction of the metals was subsequently confirmed using x-ray absorption spectroscopy. Combining physical, electrochemical, surface, and bulk properties, we proposed the following reaction mechanisms for γ-Mo_2N and VN in aqueous media, which are in good agreement with those previously reported [3].

Mo2αN+ 2H++ 4e- ↔ Mo2α-1N(H)2 $Mo_2^{\alpha}N + 2H^- + 4e^- \leftrightarrow Mo_2^{\alpha-1}N(H)_2$

VN+ 2OH- ↔ Vα-1N(OH)2 + e- $VN + 2OH^- \leftrightarrow V^{\alpha-1}N(OH)_2 + e^-$

During charge storage, application of an electric field induces insertion of H^+ and extraction of OH^- in/from the pores of γ-Mo_2N and VN materials, respectively (Fig. 1c and 1d). There was minimal contribution from large pores (mesopores and macropores) to the pseudocapacitive storage. Conversely, significant amount of H^+ and OH^- were located in small pores (~ 2 nm, upper limit for micropores) indicating that the origin of the high pseudocapacitances observed for γ-Mo_2N and VN is from the insertion/extraction of H^+ and OH^- in small pores. In other words, small pores (micropores) are the key players in the pseudocapacitive charge storage mechanisms for early transition-metal carbides and nitrides.

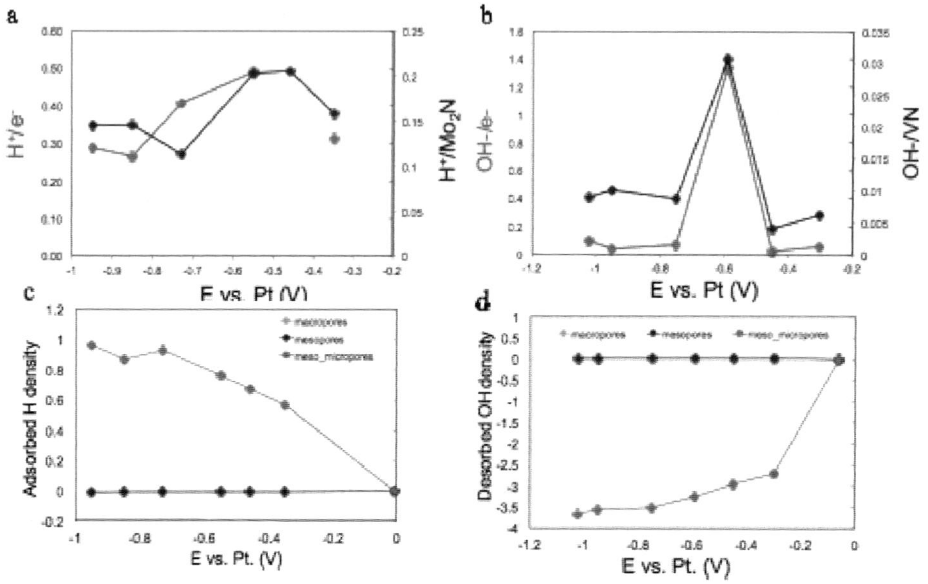

Figure 1 Hydrogen (a,c) and hydroxide (b,d) insertion and location, respectively, as function of applied potential.

References

(1) Simon, P., and Gogotsi, Y., *Nature Mat.*, **2008**, 7, 845-854

(2) Thompson, L., et al., *J. Power Sources,* **2012**, 207, 212-215.

(3) Thompson, L., et al., *J. Power Sources,* **2015**, in Press.

427199 Micro Structural and Electrical Properties of Ceria-Yttria Stabilized Zirconia Nanocomposites

Abstract:

Yttria stabilized zirconia (YSZ) is widely accepted nanomaterial electrolyte for solid oxide fuel cells (SOFCs). But, high operational temperature (~1000°C) is required in 8 mol. % YSZ (8YSZ) electrolytes for achieving acceptable levels of ionic conductivity (~0.1S/cm). Alternately, doped cerium-oxide (ceria) show promise as electrolytes capable of operating at reduced temperatures (~600°C). Enhanced ionic conductive interface is observed in doped ceria and YSZ composites in comparison to that of YSZ alone. Thus, it would be advantageous to synthesize ceria-8YSZ composites nanomaterials with 0, 5 and 10 wt. % ceria addition in 8YSZ, for attaining enhanced ionic conductivity. The novelty of current proposal is to synthesize ceria-8YSZ nanocomposite via spark plasma sintering (SPS, 1200 and 1300°C, 5min. 30 MPa), and compare the resulting properties (mechanical properties, ionic conduction, etc) of ceria-8YSZ composites. This will allow correlation of processing temperature, microstructure, and ionic conduction of ceria-8YSZ, so that a mechanism for enhanced ionic conductivity can be explained. XRD analysis revealed that all samples exhibit both cubic and tetragonal structuressynthesized via SPS. The impedance spectroscopy analysis shows that ionic conductivity isincreased by the increasing CeO_2content in 8YSZ (which gave the clear evidence for ionic conduction not electronic as ceria content increases); this is also attributed to the nano size effect, high ionic conducting pathways in dense CeO_2-YSZ nanocomposite to promoting fast oxygen ion transportation. Thus optimal selection (physical, micro structural, mechanical and ionic conduction) of ceria-YSZ electrolytes will strongly assist in designing solid oxide fuel cells with enhanced performance.

427592 Nanostructured Robust Cobalt Alloy Based Anode Electro-Catalysts with Superior Electrochemical Activity for Proton Exchange Membrane Fuel Cells

Wednesday, November 11, 2015: 8:48 AM
251F (Salt Palace Convention Center)

Prasad P. Patel[1], Moni Kanchan Datta[2,3], Oleg Velikokhatnyi[2,3], Prashanth Jampani[2] and Prashant N. Kumta[1,2,3,4,5], (1)Department of Chemical & Petroleum Engineering, University of Pittsburgh, Pittsburgh, PA, (2)Department of Bioengineering, University of Pittsburgh, Pittsburgh, PA, (3)Center for Complex Engineered Multifunctional Materials, University of Pittsburgh, Pittsburgh, PA, (4)Department of Mechanical Engineering and Materials Science, University of Pittsburgh, Pittsburgh, PA, (5)McGowan Institute of Regenerative Medicine, University of Pittsburgh, Pittsburgh, PA

The rapid depletion of fossil fuels and increased environmental pollution due to colossal consumption of ubiquitous fossil-fuel is a major impetus for efficient use of energy and exploration of renewable and clean energy sources.[1, 2] Generation of electricity from renewable energy sources, such as solar, wind without producing carbon dioxide-an undesirable green house pollutant, offers enormous potential for meeting the future energy demands.[3] Development of novel and efficient technology to store electrical energy is thus extremely important for meeting the global energy demand and environmental concerns. Fuel cell technology has garnered increasing attention over the years as it provides promising and sustainable approach for the production of continuous power with potential for reduced greenhouse gas emissions and higher efficiencies compared to hitherto combustion based technologies. In particular, proton exchange membrane fuel cells (PEMFCs) are considered to be suitable power sources for automobiles, consumer electronic devices and auxiliary power units due to the advantages of using hydrogen as fuel which is light-weight, clean and has a low operating temperature. Hydrogen also offers quick start-up, extended durability of system components, high power density with low weight and volume due to elimination of additional steps needed for fuel reformation. The simple system design would be reflected as an ease in operation, reduced cost and high reliability. However, capital cost of the system is a major constraint limiting commercialization of PEMFCs due to use of expensive noble metal based Pt/C catalyst. Hence, development of non-noble metal based catalysts with high electrochemical activity and durability is of specific interest to replace Pt/C and thus, lower the cost of PEMFC system.

The present study explores $Co_{1-x}(Ir_x)$ (x=0.2, 0.3, 0.4) solid solution alloys as anode electro-catalyst for hydrogen oxidation reaction (HOR) for PEMFC applications. **Fig. 1** shows the SEM micrograph of $Co_{0.7}(Ir_{0.3})$ with x-ray mapping of Co and Ir. The x-ray mapping shows homogeneous distribution of Co and Ir without segregation on any specific site. Electrochemical characterization has been carried out in H_2 saturated 0.5 M sulfuric acid (H_2SO_4) as an electrolyte, employing Pt wire as the counter electrode and Hg/Hg_2SO_4 as the reference electrode (+0.65 V with respect to normal hydrogen electrode, NHE), using a scan rate of 10 mV/sec at a temperature of 40°C. The solid solution electro-catalyst $Co_{1-x}(Ir_x)$ (x=0.3, 0.4) exhibit superior electrochemical activity than Pt/C with electrochemical stability matching to that of Pt/C. In addition, to obtain a better understanding of the fundamental electrochemical activity of $Co_{1-x}(Ir_x)$ electro-catalyst, first-principles calculations of the total energies and electronic structures of the model systems with chemical compositions similar to those of the experimentally synthesized materials have been carried out to complement the present experimental study.

The electrochemical study conducted in half-cell configuration and single PEMFC full cell results shows the potential of $Co_{1-x}(Ir_x)$ (x=0.3, 0.4) as replacement of Pt/C, owing to the excellent electrochemical performance and stability. These results portend significant reduction in the overall capital cost of PEMFCs. The results of structural characterization and electrochemical activity of these electro-catalysts will be presented and discussed.

References

1. C. Liu, F. Li, L.-P. Ma and H.-M. Cheng, *Advanced Materials*, 2010, 22, E28-E62.

2. J. R. Miller, *Science*, 2012, 335, 1312-1313.

3. M. S. Whittingham, *Mrs Bull*, 2008, 33, 411-419.

Acknowledgement

Research supported by the CBET, National Science Foundation, Grant-NSF-0933141. PNK also acknowledges the Edward R. Weidlein Chair Professorship funds and the Center for Complex Engineered Multifunctional Materials (CCEMM) for partial support of this research.

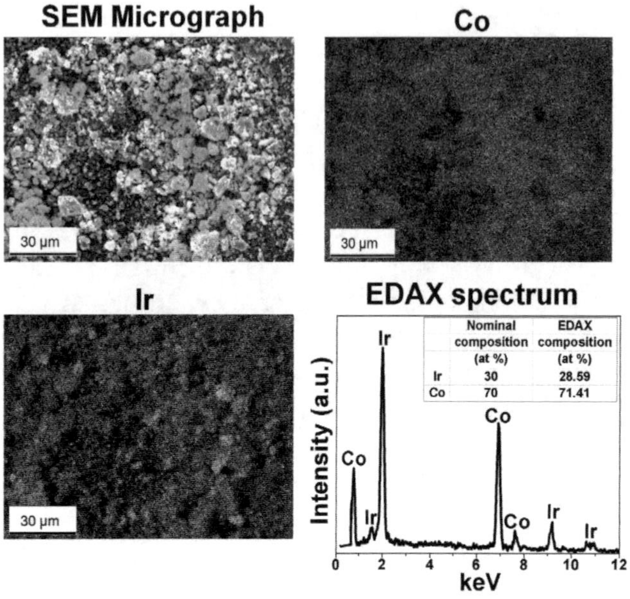

Fig 1: SEM micrograph with EDAX spectrum and x-ray mapping of $Co_{0.7}(Ir_{0.3})$ electro-catalyst

429924 Pathways Towards Defect-Tolerate, Electrochemically Stable Solar-Hydrogen Membranes

Abstract:

Defect tolerance in electrochemical corrosion protection is an important concept in chemical engineering devices. In particular, this concept has been applied to artificial photosynthetic devices that mimics nature's photosynthesis and use semiconductors to take sunlight and splits water into H_2 and O_2. Previously, more than half a century of photoelectrochemical (PEC) water splitting research mainly stayed at model systems of semiconductor – non-aqueous electrolyte interfaces, because of oxidative instability of essentially all technologically important small band gap semiconductors.

In this work, a general corrosion protection strategy for stabilizing semiconductor–electrolyte interfaces for water oxidation and hydrogen production in pH 14 base will be presented. This has been achieved by introducing complex metal-oxide-semiconductor interfaces from opto-electronics research to photoelectrochemistry. This not only brings back an entire class of semiconductors that were previously rejected, but also extends PEC theory by introducing new electrochemical models and band diagrams of stabilized semiconductor-electrolyte interfaces. Furthermore, nanowire-shape semiconductors with conformal protective coatings showed unprecedented > 2200 hours of stability for continuous water oxidation. This discrete nanowire geometry not only enables manufacture of flexible, nanowire-embedded solar-hydrogen membranes, but also confine defects within the affected region by virtue of self-passivating mechanism under the discrete nanowire geometry. Finally, I will give an outlook of the materials challenges and opportunities towards scalably manufacturable artificial photosynthetic devices that directly store solar energy into chemical fuels.

433837 PtCo/Coox Nanocomposites As Bifunctional Electrocatalysts for Oxygen Reduction and Evolution Reactions Synthesized Via Tandem Laser Ablation Synthesis in Solution-Galvanic Replacement Reactions

Abstract:

PtCo/CoO_x Nanocomposites as Bifunctional Electrocatalysts for Oxygen Reduction and Evolution Reactions synthesized via Tandem Laser Ablation Synthesis in Solution-Galvanic Replacement Reactions

Sheng Hu[2, 3], Gabriel Goenaga[2], Thomas A. Zawodzinski[2, 3], Dibyendu Mukherjee[1, 2, 3] *

[1]Department of Mechanical, Aerospace, and Biomedical Engineering; [2]Department of Chemical and Biomolecular Engineering; [3]Sustainable Energy Education and Research Center (SEERC); University of Tennessee, Knoxville, Tennessee, 37996

Intermetallic nanostructures extend the range of bulk metallic properties by virtue of their manifold compositional, structural, phase, and morphological variations. The ability to tune structure-property characteristics of metal/intermetallic nanoparticles (NPs), via controlled synthesis techniques, can enable unique physico-chemical properties that find wide-spread scientific applications in new classes of energetic and catalytic materials. In the field of proton exchange membrane fuel cell (PEMFC), efficient yet, low-cost electrocatalysts are indispensable for electrochemical oxygen reduction reactions (ORR) and oxygen evolution reactions (OER). We present a one-step, one-pot solution-phase experimental route that, for the first time, incorporates laser ablation synthesis in solution in tandem with galvanic replacement reaction (LASiS-GRR) to synthesize nanocomposites (NCs) of PtCo nanaoalloy (NA) embedded in CoO_x matrices, which act as bifunctional electrocatalysts. High resolution TEM reveals PtCo NAs of mean sizes ~8.5-17.7 nm embedded in sponge-shaped CoO_x matrices. Electron diffraction and X-ray diffraction data along with energy dispersive X-ray spectrometry (EDX) confirm the NC compositions and crystalline structures. Detailed electrochemistry data indicates outstanding ORR and OER activities for the PtCo/CoO_x NCs while exhibiting better stability than the respective standard nanocatalysts. Such activities are ascribed to the shrunken lattice constants of alloyed PtCo that promote oxygen adsorption and the synergic "spillover" effect from high surface area CoO_x matrices that accelerate both ORR and OER through symbiotic adsorption/desorption of intermediate species. Furthermore, the CoO_x matrix prevents aggregation and/or dissolution of the embedded NAs in alkaline medium. We report a combined overpotential of 756 mV vs. RHE for the PtCo/CoO_x NC with 33.3 % (molar) Pt content which is the best value ever reported for catalysts supported by carbon black. The enhanced bifunctional catalytic activities of the NCs are attributed to their unique heteronanostructures tailored by our one-pot, "green" synthesis route of LASiS-GRR.

435861 Activity of Electroless Deposited Transition Metal-Catalyst for the Alcoholysis of Ammonia Borane

Abstract:

Catalysts based on electroless nickel and bi-metallic nickel-cobalt nanoparticles were synthesized for the alcoholysis of ammonia borane for hydrogen generation. The catalysts were synthesized by polymer-stabilized Pd nanoparticle-catalyzation and activation of TiO2 substrate and electroless Ni or Co-Ni plating of the substrate for selected time lengths. Catalytic activity of the synthesized catalysts was tested for the alcoholysis of ammonia borane (NH_3BH_3) for hydrogen generation. The effects of electroless plating time lengths, temperature and NH_3BH_3 concentration on hydrogen generation rates were analyzed and discussed. Compositional analysis and surface morphology were carried out for nano-metallized TiO2 using Scanning Electron Micrographs (SEM) and Energy Dispersive X-Ray Microanalysis (EDAX). The activity of the electroless-deposited transition metal-catalysts - both under stirred and unstirred conditions - will be analyzed and discussed. The kinetics and the activation energy of the hydrogen generation alcoholysis using the catalyst will be presented.

MONOCLINIC WO_3 NANOMULTILAYERS WITH PREFERENTIALLY EXPOSED (002) FACETS FOR PHOTOELECTROCHEMICAL WATER SPLITTING

Jijie Zhang[1], Peng Zhang[1], Tuo Wang[1] and Jinlong Gong[1,]*

[1]School of Chemical Engineering and Technology, Tianjin University; Collaborative Innovation Center of Chemical Science and Engineering, Tianjin, 300072
** jlgong@tju.edu.cn*

Hydrogen is one of the most attractive and environmentally friendly fuels for energy applications [1]. Solar water splitting is a sustainable method for hydrogen generation from renewable energy resources. Photoelectrochemical (PEC) water splitting using semiconductors as photoelectrodes is a scalable and cost-effective approach. The use of tungsten oxide (WO_3) for photocatalytic reaction was first reported in 1976 [2]. Compared to benchmark TiO_2 that only responses to UV light, WO_3—a n-type semiconductor with a band gap of 2.7-2.8 eV—has been extensively investigated for water splitting under visible light irradiation. Additionally, WO_3 exhibits high resistance against photocorrosion in acidic environment under solar illumination.

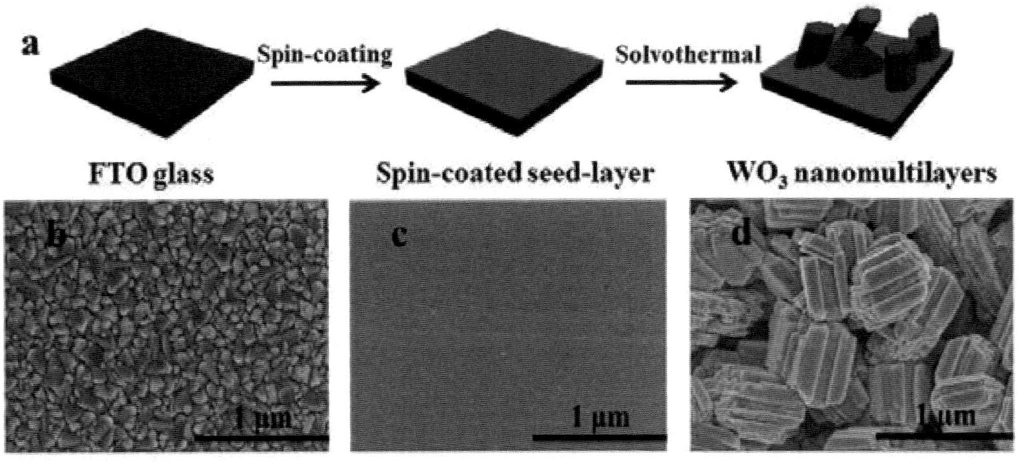

Fig. 1. (a) Schematic of the synthesis process of the WO_3 nanomultilayers, Field-emission scanning electron microscopy (FESEM) images of (b) FTO glass, (c) seed-layer, and (d) WO_3 nanomultilayers.

This work focuses on the synthesis of WO_3 nanomultilayers via a solvothermal method with ethylene glycol as the solvent, resulting in a unique multilayered structure that is different in morphology from those reported previously. A WO_3 seed-layer (Fig. 1c) was prepared by spin-coating of H_2WO_4 solution onto FTO glass followed by annealing. In each WO_3

nanomultilayered assembly (Fig. 1d), a larger backbone nanosheet is sandwiched by smaller nanosheets from both sides. For comparison, WO_3 nanorods were also synthesized as reference samples with the method reported previously.

Fig. 2a shows the photocurrent of WO_3 nanorods is 0.50 mA cm^{-2} and for the WO_3 nanomultilayers, the photocurrent density is 1.62 mA cm^{-2} at a potential of 1.25 V (*vs.* Ag/AgCl) in a 0.1 M Na_2SO_4 aqueous electrolyte (pH 6.8) under AM 1.5G irradiation. Compared with WO_3 nanorods, the WO_3 nanomultilayers exhibit an IPCE higher than 20% at 400 nm, and the IPCE increases up to 40% at 350 nm at 0.67 V *vs.* Ag/AgCl (1.27 V *vs.* RHE) in 0.1 M Na_2SO_4 aqueous electrolyte (pH 6.8) (Fig. 2b). According to the Mott-Schottky plots at 1000 Hz in a 0.1 M Na_2SO_4 solution (Fig. 2c), the WO_3 nanomultilayers show a substantially smaller slope compared to the WO_3 nanorods, suggesting a faster charge transfer than the nanorod sample. The I-t plot under AM 1.5G at a constant applied bias of 1.0 V (*vs.* Ag/AgCl) in a 0.5 M H_2SO_4 electrolyte solution (pH 0.3) (Fig. 2d) shows that the photocurrents of WO_3 nanomultilayers and WO_3 nanorods remain stable at 0.5 and 0.2 mA cm^{-2}, respectively, over a 4-h period. From these characterization, it was absolutely seen the WO_3 nanomultilayers have better PEC performances than WO_3 nanorods.

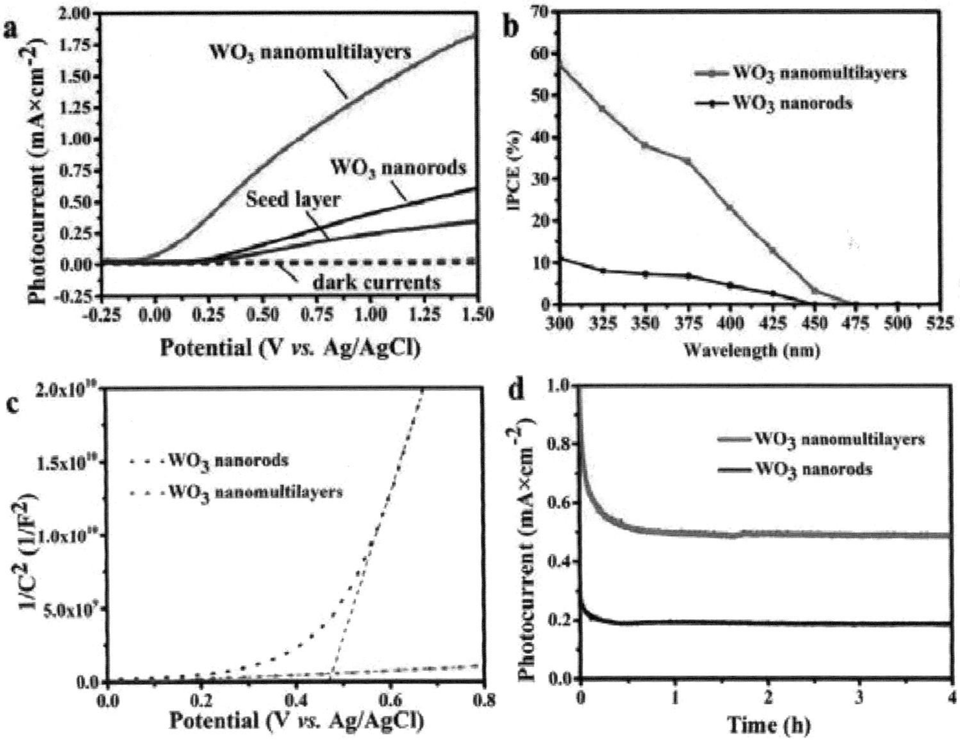

Fig. 2. (a) Current-potential plots of WO_3 nanomultilayers, WO_3 nanorods and WO_3 seed-layer; (b) IPCE plots, (c) Mott-Schottky plots and (d) photocurrent-time plots of WO_3 nanomultilayers and WO_3 nanorods.

In order to explore the origin of the enhanced photocatalytic activity, X-ray diffraction (XRD) patterns of FTO glass, WO_3 seed-layer, WO_3 nanorods and WO_3 nanomultilayers were obtained (Fig. 3a). According to the XRD patterns of the two samples, the (002), (020) and

(200) planes are exposed in γ-monoclinic WO_3. Among the common crystal phases of WO_3, γ-monoclinic is the most stable phase at room temperature. In this study, the WO_3 nanomultilayers with (002) facets preferentially exposed were synthesized by using ethylene glycol solvent, as confirmed from the XRD patterns. It is well-known that the photocatalytic reactivity of a semiconductor photocatalyst is affected by its electronic and surface atomic structures, which is sensitive to the crystal facet with different orientations. WO_3 nanostructures with different crystal planes possess surface energies following an order of (002) (1.56 J m^{-2}) > (020) (1.54 J m^{-2}) > (200) (1.43 J m^{-2}) [3]. It has been reported that the preferential orientation of (002) planes of WO_3 was possibly more favourable in absorption and redox of pollutants than preferential orientation of (020) planes [4]. The preferential exposure of the highly active (001) facets, facets that belong to the same {001} family of planes as (002) facets, endows the WO_3 nanocrystals with significantly enhanced photocatalytic activity, which can be attributed to the effective production of active oxygen species and reduced recombination of photogenerated electrons and holes. The (002) facets are the most active ones among the three facets of WO_3 and considered to have superior reactivity in nanoparticulate photocatalysis and PEC reactions. The outstanding photocurrent and IPCE results suggest that the preferential exposure of (002) facets of WO_3 nanomultilayers could effectively improve the PEC water splitting performance of WO_3.

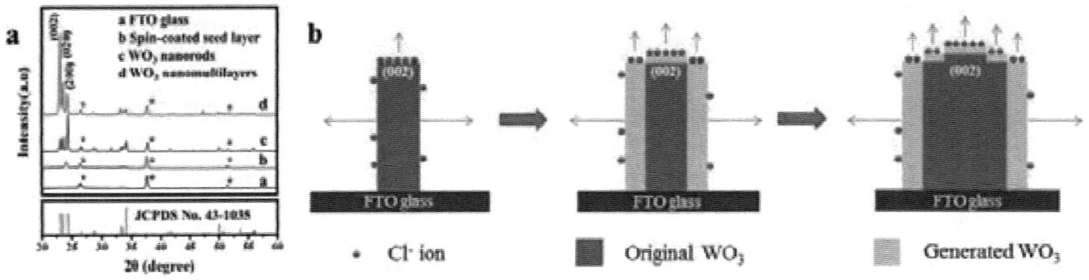

Fig. 3. (a) XRD patterns of WO_3 nanomultilayers, WO_3 nanorods, spin-coated seed-layer and FTO glass. Peaks marked with asterisk originate from the FTO substrate. Vertical bars indicate peak position and intensity of monoclinic WO_3 (JCPDS No. 43-1035); (b) Schematic of the crystal growth process of the WO_3 nanomultilayers. The direction of the arrow stands for the growth direction of different facets and the length of the arrow stands for the growth speed of different facets.

We propose a possible growth mechanism of the WO_3 nanomultilayers. During the solvothermal reaction, the Cl^- functioned as a kind of capping ion like F^- which is reported previously [5,6]. Since the (002) facets have lager surface energy than others, Cl^- is preferably adsorbed on the (002) facets. This absorption will weaken the surface affinity for the precursor to this orientation, thus slow down the growing rate along this orientation, and resulting in the increased exposure of the (002) facets. During the anisotropic growth, this difference of growth rate and capping effect will result in step-like morphology. The process of crystal growth is schematically illuminated in Fig. 3b. While in the hydrothermal reaction of WO_3 nanorods sample preparation, the Cl^- ions in water have less electronegative than in ethylene glycol, which results in the adsorption of less Cl^- on the (002) facets. Therefore few

(002) facets survived after coarsening.

In summary, the γ-monoclinic WO_3 nanomultilayers with highly reactive (002) facets exposed have been synthesized by a solvothermal method using ethylene glycol as the solvent. WO_3 nanomultilayers exhibit an excellent photocurrent of 1.62 mA cm^{-2} at 1.25 V (*vs.* Ag/AgCl) under AM 1.5G irradiation, and an IPCE of 40% at 400 nm at a bias of 0.67 V (*vs.* Ag/AgCl) due to their unique multilayered nanostructure and the preferential exposure of highly reactive (002) facets. WO_3 with well controlled nanostructure and (002) facets preferentially exposed is confirmed to exhibit enhanced PEC performance, which will hopefully stimulate more extensive and insightful investigations of WO_3 nanostructures as PEC photoanodes.

Reference:

1. J. Graetz (2009), "New approaches to hydrogen storage," Chem. Soc. Rev., 2009, 38, 73-82.

2. G. Hodes, D. Cahen and J. Manassen (1976), "Tungsten trioxide as a photoanode for a photoelectrochemical cell (PEC)", Nature, 1976, 260, 312.

3. D. Zhang, S. Wang, J. Zhu, H. Li and Y. Lu (2012), "WO_3 nanocrystals with tunable percentage of (001)-facet exposure", Appl. Catal., B, 2012, 123-124, 398-404.

4. Y. Guo, X. Quan, N. Lu, H. Zhao and S. Chen (2007), "High photocatalytic capability of self-assembled nanoporous WO_3 with preferential orientation of (002) planes", Environ. Sci. Technol., 2007, 41, 4422-4427.

5. H. G. Yang, C. H. Sun, S. Z. Qiao, J. Zou, G. Liu, S. C. Smith, H. M. Cheng and G. Q. Lu (2008), "Anatase TiO_2 single crystals with a large percentage of reactive facets", Nature, 2008, 453, 638-641.

6. X. W. Zhao, W. Z. Jin, J. G. Cai, J. F. Ye, Z. H. Li, Y. R. Ma, J. L. Xie and L. M. Qi (2011), "Shape- and Size-Controlled Synthesis of Uniform Anatase TiO_2 Nanocuboids Enclosed by Active {100} and {001} Facets", Adv. Funct. Mater., 2011, 21, 3554-3563.

Transparent ALD-grown Ta_2O_5 protective layers for corrosion vulnerable photoanodes in solar water splitting

Tuo Wang*, Chengcheng Li, Jinglong Gong

Key Laboratory for Green Chemical Technology of Ministry of Education; School of Chemical Engineering and Technology, Tianjin University; Collaborative Innovation Center of Chemical Science and Engineering; Tianjin 300072, China

*E-mail: wangtuo@tju.edu.cn

Photoelectrochemical water splitting for hydrogen production by semiconductors has been extensively investigated in recent decades for the conversion and storage of the intermittent solar energy. However, photocorrosion caused by oxidative holes has restricted the application of many promising semiconductors (such as ZnO, Si) during the water oxidation reaction, which is considered particularly demanding compared with the reduction of protons. Chemically stable protective layers have been investigated to alleviate this problem, whereas their optical properties are rarely examined explicitly as a contributing factor. This work describes the design and fabrication of highly stable corrosion vulnerable photoanodes with ultrathin (~1.5 nm) Ta_2O_5 protective layers conformally deposited by atomic layer deposition. The photocurrent of the protected electrode remains stable during stability tests over several hours without degradation, while the bare electrode shows a fast decay because of photocorrosion. The optical property of Ta_2O_5 ensures its transparency to sunlight, prohibiting the generation of high energy oxidative holes from Ta_2O_5 that will attack the underlying material. The stability of protected electrode is deteriorated upon the incorporation of high energy UV light (¦Ë < 290 nm) to deliberately generate holes from Ta_2O_5. It is concluded that the transparency of Ta_2O_5 to sunlight is the main reason accounting for the excellent stability of the photoanode. Moreover, the photoactivity of the electrode is improved due to the surface state passivation effect of Ta_2O_5, and the photocurrent can be further increased upon the deposition of additional surface catalytic layers such as NiO. This study may provide a new perspective on selecting protective materials for unstable semiconductors by evaluating the transparency of the overlayer.

430533 Bimetallic Cu/Pd Nanoparticles As Low Temperature Sulfur-Tolerant WGS Catalysts

Abstract:

Bimetallic Cu/Pd nanoparticles as low temperature sulfur-tolerant WGS catalysts

SeongUk Yun and Vadim V. Guliants

Chemical Engineering, University of Cincinnati, Cincinnati OH 45221

The global interest in the water-gas shift (WGS) reaction as a method of making hydrogen from coal-gasification products increased significantly in recent years. However, typical low temperature WGS catalysts are easily deactivated by sulfur compounds present in coal-derived syngas. According to recent theoretical predictions, Cu and Pd are expected to be promising sulfur-tolerant WGS catalysts. In addition, bimetallic Cu/Pd nanoparticles exhibiting bulk alloy, surface alloy, and core-shell structures are expected to show differences not only in the WGS activity, but also tolerance to sulfur.

Several bimetallic Cu/Pd catalysts were synthesized in our laboratory and characterized by scanning electron microscopy (SEM), transmission electron microscopy (TEM), X-ray diffraction (XRD), thermogravimetric analysis (TGA), physi/chemisorption, and temperature-programmed reduction (TPR) in H_2. The TPR analysis showed that the presence of Cu facilitated the reduction of active metals at lower temperatures. The TGA analysis revealed a strong interaction between Cu, Pd and oxide support suggesting solid alloy information. The XRD indicated that the size of these metallic nanoparticles was in the 15-20 nm range depending on the Cu/Pd ratio. The size and shape of supported Cu/Pd nanoparticles were determined by TEM. The WGS catalytic activity of catalysts have been evaluated in the WGS reaction at 398-723 K under atmospheric pressure in a fixed-bed microreactor employing the feed of 10% CO and 20% H_2O (balance He). Significant improvement in the WGS catalytic activity was observed by optimizing the Cu/Pd ratio. These novel bimetallic catalysts showed improved sulfur resistance during WGS reaction.

432298 N2/Ar-Plasma Assisted Nitrogen Doping of Ordered Mesoporous TiO2 Thin Films for Water Splitting Photocatalysis

Abstract:

TiO_2 has attracted much attention due to many advantageous properties including its low cost, high availability, chemical stability, and excellent opto-electronic properties. These unique properties have enabled titania to be utilized in a wide range of applications including photocatalysts, photovoltaics and photoconductors for energy production and environmental remediation. Despite many attractive features of TiO_2, one critical challenge is the innate inability of TiO_2 to absorb visible light. In this study, this shortcoming is addressed by doping cubic ordered mesoporous TiO_2 thin films (N-TiO_2) using nitrogen/argon (N_2/Ar) plasmas. This represents a rapid, scalable approach to high surface, controlled nanostructure thin films for photocatalytic applications. For this purpose, cubic ordered mesoporous TiO_2 thin films were prepared by a surfactant templated sol-gel method using $TiCl_4$ as precursor and triblock copolymer Pluronic F-127 as the template. Subsequently, the calcined TiO_2 films were treated with N_2/Ar plasma under controlled conditions of reaction gas pressures, microwave power, and plasma exposure duration. The primary variable studied here for N-TiO_2 films is the duration of plasma exposure (0-210 min). After doping, the films were confirmed to have an accessible, well-ordered cubic mesopore structure by SEM, TEM, STEM and XRD analyses. The content of nitrogen in the films increases with plasma exposure duration, up to over 3% atomic N. X-ray photoelectron spectroscopic (XPS) analyses and UV-vis absorbance spectra of N-TiO_2 films indicate that the incorporated N atoms reduce the band gap of TiO_2 and thus enhance the absorption of visible light. Finally, the visible-light photocatalytic activity of N-TiO_2 films was determined from the photocatalytic degradation of methylene blue under illumination with a visible-light LED (455 nm wavelength). The N-TiO_2 films prepared with 150 min plasma treatment show the best photocatalytic activity, with a first-order methylene blue decolorization rate coefficient 6 times greater than that of undoped TiO_2 films. N-TiO_2 films processed under further optimized conditions also showed about 80 times greater photocurrent in visible-light photoelectrochemical water splitting than undoped TiO_2 films. Insights from electrochemical impedance spectroscopy into the electronic structure and conductivity of the N-TiO_2 films will be discussed.

435770 TiO2/BaTiO3 Bi-Layered Approach Towards Photoelectrochemical Generation of Hydrogen Via Direct Splitting of Water

Abstract:

TiO_2/BaTiO3 bi-layered approach towards photoelectrochemical generation of hydrogen via direct splitting of water

Shailja Sharma, Gurpreet Kaur, Anuradha Verma, Anamika Banerjee, Vibha R Satsangi[1], Sahab Dass, Rohit Shrivastav*

Department of Chemistry, Dayalbagh Educational Institute,

Dayalbagh, Agra-282 005, India

[1]Department of Physics & Computer Science, Dayalbagh Educational Institute,

Dayalbagh, Agra-282 005, India

*Author for correspondence: rohitshrivastav_dei@yahoo.co.in

Abstract:

Nanostructured bi-layered thin films have potential application as photoelectrode in photoelectrochemical (PEC) cell for generation of hydrogen by splitting of water. This study involves the fabrication of bi-layered junction which explored for best performance by tailoring thickness of the $BaTiO_3$ over TiO_2 thin films. Pristine $BaTiO_3$, TiO_2 and $BaTiO_3$/ TiO_2 were synthesized by sol- gel spin coating technique and were subjected to XRD (for phase and particle size analysis), UV-spectroscopy (for optical characterization), AFM (for surface topography) and SEM (for surface morphology). Wide absorption spectrum in visible region along with a sharp shoulder in UV region was observed. AFM and SEM picture of bi-layered thin films revealed uniform and well defined distribution of nanoparticles on the substrate i.e. growth of crystallites perpendicular to the substrate. XRD analysis confirmed the exhaustive evolution of tetragonal $BaTiO_3$ and anatase TiO_2 in all the samples, particle size varied between (40-50 nm). A maximum photocurrent density of ~0.98 mA/cm ^{2} at 0.9 V/SCE was recorded for 674 nm thick modified TiO_2/$BaTiO_3$ bi-layered photoelectrode, results are encouraging and show that bi-layered films yielded significant gain in photocurrent compared to mono-layered pristine samples.

Keywords: TiO_2/$BaTiO_3$ bi-layered thin films, Photoelectrochemical cell, solar hydrogen production, Water Splitting.

AUTHOR INDEX

Agrawal, Rakesh 6
Aksoy, Burak 93
Albano, Fabio 85
Alhosani, Abdullah 34
Amama, Placidus B. 83
Apblett, Christopher A. 65
Arnold, Robert 31
Baah, David 100
Balani, Kantesh 119
Bardhan, Rizia 4
Barrett, Lawrence 68
Barringer, Dave A. 65
Berger, Jordan 45
Birkmire, Robert W. 7
Booshehri, Amin Yoosefi 56
Boyne, Robert W. 6
Brett, Dan 114
Brown, Paul 94
Brunschwig, Bruce S. 123
Brushett, Fikile 67
Buonsanti, Raffaella 58
Cai, Tianhao 2
Carpena, Jennifer 83
Chatterjee, Anushree 28
Chen, Rongjun 32
Chen, Weixiang 81
Cheng, Yang -Tse 84
Choi, Saemin 116
Chou, Chiayun 78
Cliffel, David 55
Collinson, Maryanne M 93
Coppens, Marc-Olivier 114
Courtney, Colleen 28
Cullinan, Harry T. 93
Dagnall, Katelyn 49
Dahlberg, Kevin 85
Dames, Chris 58
Damle, Sameer 61
Dapkus, P. Daniel 123
Darling, Robert 67
Datta, Moni Kanchan 120
David, Allan 31
Davis, Robert C. 68
Demir, Muslum 93
Deng, Da 79
Dhar, Subhash 85
Ding, Bo 81
Ding, Yuchen 115
Djire, Abdoulaye 116
Dong, Lianyang 60
Durstock, Michael F. 83
Efeovbokhan, Vincent 125
El-kaderi, Hani M 93
Fairbairn, Heather 50
Fan, Juichin 68
Farghaly, Ahmed A. 93
Fei, Ling 82
Fenton, Kyle R. 65
Follette, Marissa 83
Forster, Jason 58
Gallagher, Kevin 67
Gardner, David W. 35
George, Steven 96
Ghosh, Swapnadip 2
Gizzie, Evan 55
Gomes, Margarida Costa 94
Gong, Jinlong 126, 130
Goodman, Samuel 28, 54
Gorrell, Paul 100
Grillet, Anne M. 65
Guliants, Vadim V. 131
Guo, Juchen 64
Guo, Zhanhu 66
Gupta, Alka 119
Gupta, Ram B. 93
Hamdeh, Umar H. 1
Han, Kai 98
Han, Sang Eon 2, 51
Han, Sang M 2
Han, Seok Jun 2
Harb, John 27, 68
Hare, Christopher D. 96
Hart, Noam 64
Hatton, T. Alan 94
He, Lilin 116
Hoard, Brittany R. 2
Hong, Joung Sook 30
Hu, Sheng 124
Hu, Shu 123
Huffman, Daniel R. 83
Humplik, Thomas 65
Hwang, Gyeong 59, 78, 97
Islam, Syed Z. 84, 132
Islamoglu, Timur 93
Jampani, Prashanth 120
Jennings, G. Kane 55
Ji, Ge 81
Ji, Shaowen 30
Jiang, Guo 66
Johnson, Donald 100
Joo, Yong L. 82, 99
Kalu, Egwu E. 125
Kang, Yongmin 57
Kelly, Alexander 31
Kim, Doo Young 132
Kim, Jangwoo 99
King, David 85
Kumta, Prashant N. 61, 120
Kwee, Darren W. 77
Lawlor, Colleen C. 47

AUTHOR INDEX

Lee, Ilsoon 30
Lee, Jim Yang 81
Lee, Myongjai 85
Lee, Yongjin 59
Lewis, Nathan S. 123
Li, Chengcheng 130
Lichterman, Michael 123
Liu, Miao 58
Long, Kevin N. 65
Lovelett, Robert J. 7
Lu, Meihua 81
Lu, Nancy 33
Mahajan, Vishal 85
Maiti, Spandan 61
Mao, Xianwen 94
Martin, C. 46
Martinez-Morales, Alfredo A. 77
Maruyama, Benji 83
Mbah, Jonathan C. 100
Mendoza, Hector 65
Milliron, Delia J. 3
Miskin, Caleb 6
Mohanty, Debasish 85
Mohraz, Ali 113
Mumm, Daniel R. 113
Musgrave, Charles B. 96
Nagpal, Prashant 28, 52, 54, 115
Nagpure, Suraj 84
Nelson, Rainie D. 1
Neuber, Markus 96
Neville, Tobias 114
Ng, K. Y. Simon 79
Nuraje, Nurxat 5
Ogunnaike, Babatunde A. 7
Oguntoye, Moses 95
Omoleye, James A. 125
Ong, Gary K. 3
Onyeozili, Edith O. 125
Oviedo, M. Belen 53
Pak, Alexander 59, 97
Pal, Siladitya 61
Palmer, Sydney 68
Panthani, Matthew G. 1
Patel, Prasad P. 120
Persson, Kristin 58
Pesika, Noshir S. 95
Phapal, Sopan M 29
Qu, Baihua 81
Rajput, Nav Nidhi 58
Rankin, Stephen E. 84, 132
Rasmussen, Paul G. 116
Reed, Allen 132
Roberts, Scott A. 65
Robinson, Maxwell 55
Russ, Boris 58
Sahu, Ayaskanta 58
Schnabel, Hans-Dieter 96
Segalman, Rachel 58
Shafarman, William N. 7
Shaner, Matthew 123
Sharma, Shailja 133
Shearing, Paul 114
Siegel, Jason 116
Siegrist, Theo 60
Singh, Ajay 3
Sleightholme, Alice E. S. 116
Smith, Kurt B. 76
Smith, Samantha 36
Snyder, Chelsea 65
Song, Anna 30
Stanley, Joel 85
Stevenson, Lisa 85
Stirrup, Emily K. 65
Su, Liang 67
Sun, Ke 123
Sun, Qi-C. 52
Sunthar, P. 29
Thompson, Levi T. 116
Tomassone, M. Silvina 76
Torralba, Monica 48
Trogadas, Panagiotis 114
Uprety, Bibek 27
Urban, Jeffrey 58
Vaddiraju, Sreeram 57
Vahdat, Nader 100
Vanfleet, Richard R. 68
Vasiraju, Venkata 57
Velikokhatnyi, Oleg 120
Wang, Tuo 126, 130
Wang, Yiran 66
Ware, Ashley 100
Wei, Huige 66
Wei, Suying 66
Witt, Jessica A. 113
Wong, Bryan 53
Wood, David 85
Xiang, Chengxiang 63
Xie, Wei 67
Xu, Rong 56
Yang, Fan 58
Young, Matthias J. 96
Yun, SeongUk 131
Zhang, Jijie 126
Zhang, Qinglin 84
Zhiyuan, Xiong 86
Zhu, Jian 79